A Life of the Land:
Connecticut's Jewish Farmers

Mary M. Donohue and Briann G. Greenfield

Volume 4   Fall 2010
Connecticut Jewish History

Funded by the State of Connecticut

Connecticut Commission on Culture & Tourism

Connecticut Jewish History, Volume 4

Project Directors
Estelle Kafer, Executive Director, Jewish Historical Society of Greater Hartford
Mary M. Donohue, Architectural Historian, Commission on Culture & Tourism

Editors and Authors
Mary M. Donohue, Architectural Historian, Commission on Culture & Tourism
Briann G. Greenfield, Ph.D., Central Connecticut State University

Graphic Designer
Carla Iovinella

Copy Editor
Jennifer Huget

Printing
Pyne-Davidson Printers

Published by the Jewish Historical Society of Greater Hartford
333 Bloomfield Avenue, West Hartford, Connecticut 06117
www.jhsgh.org, 860-727-6171
Additional copies of the journal may be purchased by contacting the Society

Copyright 2010 by the Jewish Historical Society of Greater Hartford
All rights reserved. No part of this publication may be reproduced in any form or by any electronic or mechanical means including information storage and retrieval systems without permission in writing from the publisher. JHSGH and the editors of this journal disclaim responsibility for statements of fact or opinion made by contributors. Statements expressed by contributors are their own views and not necessarily the views of the Society.

ISBN: 978-0-615-28809-3

The Jewish Historical Society of Greater Hartford is a non-profit beneficiary agency of the Jewish Federation of Greater Hartford.

Publication of the journal was made possible with funding from the Connecticut Commission on Culture & Tourism with funds from the Community Investment Act of the State of Connecticut. Additional funding was received from the Feltman Family Fund and the Irving and Evelyn Gilman Fund of the Jewish Community Foundation of Greater Hartford.

Connecticut Commission
on Culture & Tourism

Front Cover: Samuel Schwartz feeding poultry on farm in Lebanon, 1946. Courtesy of the JHSGH

A LIFE OF THE LAND:
CONNECTICUT'S JEWISH FARMERS

Table of Contents

President's Message   4
Elliott B. Pollack

Executive Director's Message   5
Estelle Kafer

A Life of the Land: Connecticut's Jewish Farmers   6
Mary M. Donohue and Briann G. Greenfield

Timeline   28

Remembering a Life of the Land: An Oral History of Connecticut's Jewish Farmers   30
Briann G. Greenfield and Lucas Karmazinas

Farmers on Film: Photographs from the Collections of the Farm Securities Agency and the Office of War Information   66
Briann G. Greenfield

Faith Amidst the Fields: Connecticut's Country Synagogues   80
Mary M. Donohue and Robert Gregson

The Catskills of Connecticut: Jewish Farming Communities as Summer Retreats   92
Briann G. Greenfield

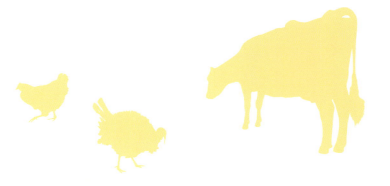

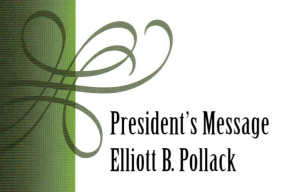

President's Message
Elliott B. Pollack

In 2010, it might at first seem a bit nostalgic, if not incongruous, for the Jewish Historical Society of Greater Hartford to sponsor a volume which studies and commemorates the lives and work of Connecticut's Jewish farmers. In fact, many readers of this volume will express surprise that such a demographic group existed!

Quite to the contrary, Jewish farm life was an important feature of early to mid-twentieth century Connecticut. Many newly arrived Jews lacking family connections or seeking relief from the increasingly lonely crowd, to pirate Reisman's term, in America's cities, or for financial reasons, made their lives in poultry and dairy farming throughout Connecticut, primarily in New London, Windham, Tolland and Hartford counties. Some farms developed into resorts that were also akin to Catskills resorts, but on a far smaller scale, catering to vacationing urbanites seeking a bigotry free rural interlude.

I trust you will agree with me, after reading the fascinating documentation of Jewish farm families' lives which has been so skillfully assembled in this memorable volume, that it would have been unthinkable for the Jewish Historical Society not to have supported this critical retrospective.

It is not relevant that the children and grandchildren of Connecticut's Jewish farmers, with few exceptions, did not emulate their forebears and continue to draw their sustenance from the land. Time, upward mobility and all of the other sociological factors of which we are so well aware, together with suburban sprawl, made that likelihood virtually impossible. Notwithstanding, their forebears' encounters with a rural Connecticut that has virtually disappeared must be celebrated and not lost to our posterity.

If it is true that our ability to understand our lives requires an understanding of what past generations and societies experienced, we are profoundly indebted to Briann G. Greenfield and Mary M. Donohue for bringing this research and analysis to our attention.

Elliott B. Pollack
President
2007-2010

Executive Director's Message
Estelle Kafer

The Jewish Historical Society of Greater Hartford began collecting the oral histories, photographs and historical documents of Connecticut's Jewish farmers in the 1980s. Their remarkable stories and accounts of starting a new life and earning a livelihood in a remote location, far removed from their families and community, demonstrates their determination, courage and resilience. The work was difficult, the benefits few, but Jewish farmers managed to establish self-sufficient communities, while adhering to their religious observances, traditions and values. Their social and economic interaction with, and contributions to the larger community were nationally recognized and had a far reaching and long lasting impact on the area's agriculture and industry.

I recall visiting a number of farmers as well as their descendants, in Lebanon, Colchester and Norwich to assist our historians, Dr. Kenneth Libo, Adjunct Professor of History, Hunter College, and Mary M. Donohue, Architectural Historian, Commission on Culture & Tourism, while I have worked on this project. I listened to recollections of hardship, endless labor and a simple but happy life. I also detected a sense of pride, dignity and accomplishment as the narrators spoke of their state of the art machinery, crops or livestock, family cohesiveness and farming challenges. This significant chapter in Connecticut history has now been documented so that researchers and future generations can understand the sacrifices and achievements of these ingenious immigrants.

On behalf of the Society, I would like to thank editors and authors Mary M. Donohue and Briann G. Greenfield for the countless hours they spent researching, writing and organizing a vast amount of historical data and material for this journal. I would also like to thank the Society staff, Bea Brodie, and Cynthia Harbeson, for their diligent assistance in the many months of preparing this book for publication. Other members of the project team consisted of Carla Iovinella, graphic designer and Jennifer Huget, copy editor, Lucas Karmazinas, CCSU graduate student, and proofreaders Barbara Gordon and Reba Nassau.

I am especially grateful to the Connecticut Commission on Culture and Tourism, which recognized the value of this research and provided funding for the publication of this journal. The Connecticut Humanities Council gave the Society an initial grant for preliminary research and we also received funding from the Feltman Family Fund and the Gilman Family Fund of the Jewish Community Foundation of Greater Hartford. I thank the Society's Board of Directors for their ongoing support and confidence that this long-term project would culminate in this important scholarly work.

Estelle Kafer

"Never buy a farm in a hurry. Never buy a farm unless you have capital enough to keep you the first year. Not every Jew is a farmer. Don't think of buying a farm if your wife does not like farm life."

Guide to the United States for the Jewish Immigrant, A Nearly Literal Translation of the Second Yiddish Edition, John Foster Carr, published by the Immigrant Publication Society of New York City, 1916.

A LIFE OF THE LAND:
Connecticut's Jewish Farmers

Mary M. Donohue and Briann G. Greenfield

Who ever heard of a Jewish farmer?

Connecticut's Jewish farmers have been considered a novelty since they began to arrive from Eastern Europe in the 1890s. After all, Jews typically had not been associated with farming—and certainly not with successful farming. But in this tiny New England state, Jews carved out lives as farmers, bringing to the experience innovations that would come to distinguish them as outstanding in the field.

Though forbidden to own land in Russia and Russian-occupied countries, Jews still came to America with some agricultural skills gained through cattle dealing, tenant farming, or raising cows, goats, or chickens. In Connecticut, Jewish farmers moved past those basics, pioneering the adoption of foodstuffs such as eggs, milk, and broilers that could be raised on worn-out, rocky New England soil. (After all, rocks are known as "Connecticut potatoes.") Tobacco cultivation also proved a moneymaker for these Jews.

Leon, Simon (on top of hay) and Samuel Schwartz with haywagon in Lebanon, ca. 1940s

Beyond specific crops, though, large-scale use of scientifically designed chicken coops, the Jewish farm agent, the Jewish farm newspaper, and the Jewish farmers' cooperative all became highly visible hallmarks of Jews' success at farming in Connecticut. Similarly, the farmers' credit union in Fairfield, formed by the Jewish Agricultural Society in 1911, was among the first in the nation to create a model that soon was adopted across the country.

Coming to Connecticut

The Pale, 1835-1917

Masses of Eastern European Jews began immigrating to the United States in the 1880s. Between 1881 and 1924, more than two-and-a-half million Jews arrived in America, the overwhelming bulk of them either directly or indirectly from Eastern Europe. Eastern European Jewish immigrants settled largely in cities of the Northeast and Midwest. New York City alone was home to an estimated one-third of all Jews living in the United States in 1880 and close to half the country's Jewish population in 1920.[1] With its proximity to New York, Connecticut developed a substantial Jewish community. By 1910, Hartford counted more than 6,500 Jews among its citizens, with approximately 5,000 having arrived from Eastern Europe within the preceding three decades.[2]

In leaving Eastern Europe, these Jewish immigrants fled nearly two centuries of persecution that had only become more pronounced after the assassination of the Russian Czar Alexander II in 1881. In Eastern Europe, the Russian government required Jews to live in the Pale of Settlement, a region stretching from the Baltic to the Black Sea. They were prohibited from owning land and often were forced to live in ghettos in urban areas,

Lower East Side, Manhattan, ca. 1900s

where they typically found work as merchants and artisans. Anti-Semitic violence was common as the Russian government condoned *pogroms,* organized raids on Jewish communities that included killings and the destruction of property. These immigrants had many reasons to leave and little incentive to return home.[3] As a *Hartford Daily Courant* reporter explained in 1882, "When asked whether these Russian Jews expect to abide permanently in this country or to return to their native land," the reply was that "they had come for good."[4] Beginning with the earliest wave of Jewish refugees that in 1881 flooded Brody, a western Ukrainian town on the Austrian border, few Jews had the resources to provide themselves with the basics: food, shelter, and clothing.[5]

Arriving in Connecticut, Eastern European Jews found an existing community of German Jews who had come in the earlier part of the nineteenth century and had settled in the state's cities, predominantly New Haven, Bridgeport, Waterbury, and Hartford. Smaller in number, German Jews had quietly established themselves in Connecticut, many finding success as businessmen and merchants. In 1843, they successfully petitioned the state legislature to amend state statutes to permit public worship by Jews and quickly formed Orthodox congregations in the state's larger cities. Beginning in the 1870s, many German Jews adopted the practices of Reform Judaism, which allowed them to assimilate more easily into their new communities.

Divided by culture, language (German Jews did not speak Yiddish), and class from the Orthodox Eastern European newcomers, members of Connecticut's German Jewish community nevertheless worked to assist their coreligionists. In 1882, they established the Society for the Aid of Russian Refugees. A brief notice appearing under the headline "The Russian Refugees" in the *Hartford Daily Courant* on June 28, 1882 announced the organization's mission and asked for donations:

> The committee of the local Russian Aid society, engaged in the relief of the Russian Jew refugees in this city, [asks for] assistance from the public. Money may be sent to the COURANT office and will be duly acknowledged in these columns, or assistance may be given by procuring employment. Hartford has received a large number of the refugees, work has been found for 200 in factories and on farms, and others arriving daily.[6]

Directed by L.B. Haas, a Hartford tobacco merchant whose family had emigrated from Holland, the Society for the Aid of Russian Refugees belonged to an international network of aid organizations that sought permanent homes for those affected by the *pogroms*. Haas himself had attended a meeting in New York City in June 1882 called by the Hebrew Immigrant Aid Society of the United States and attended by individuals representing nearly twenty-five separate relief organizations from various regions of the United States and Canada. Delegates heard accounts of relief efforts in Paris, London, Berlin, Frankfurt, and other European cities along with a report on efforts, modeled after an earlier arrangement to aid Irish immigrants, to colonize Jewish refugees on Western farms.[7] By the time of the conference, New York's Hebrew Immigrant Aid Society had relocated some two hundred Jewish immigrants to Hartford. In June, another group arrived, this one directly from Liverpool. This later group's transportation was facilitated by the British National Mansion House Committee, one of the most important committees organized in Europe to assist victims of the *pogroms* and one that used Liverpool as a base for relocating people to the United States.

In these first years, the Hartford Aid Society often found work for newcomers in factories such as the Ponenah Mills near Norwich and the Eagle Lock Company in Tariffville. But the idea of rural resettlement, which had been used by American charity workers since the 1850s as a means of dealing with the nation's urban poor, was clearly under consideration.[8] In a *Hartford Daily Courant* article, a representative of the Hartford Aid Society listed "farmers" as among the trades represented by the new immigrants and also noted the existence of cooperative farming ventures in Oregon and Louisiana. Locally, Jewish immigrants had been placed as workers on New England farms, and while some "vigorously objected to working on Saturdays and eating meat not slaughtered in accordance with Jewish rites," the Society's representative nevertheless claimed that Connecticut Jews saw agricultural as an avenue to prosperity.[9]

Baron de Hirsch

The Baron de Hirsch Fund and the Jewish Agricultural and Industrial Aid Society

The most significant support for Jewish farmers in Connecticut came from European sources. The Baron Maurice de Hirsch (1831-1896) was an extremely wealthy German Jew who amassed a fortune as an industrialist and builder of railroads. Hirsch became a strong proponent of the Jewish "Back to the Land" movement, an international effort to resettle persecuted Jews in rural colonies that stressed the redemptive nature of farm life. His only son had died as a young man, prompting the Baron to devote much of his fortune to philanthropy. In 1881, he had donated one million francs to ease the Jewish refugee situation in Brody. Convinced that Jews did not have a bright future under Russian rule, he became interested in their emigration to the United States.

Abraham Leon Greenberg (1847-1907), farmer, Farmington

As early as 1889, the Baron contacted the Alliance Israélite Universelle, a Paris-based international Jewish organization, with an offer to provide funds to assist Russian and Romanian immigrants in the United States. Isidore Loeb, the secretary of the Alliance Israelite Universelle, notified a number of prominent American Jewish leaders of the Baron's desire to have an American committee administer his relief funds. A conference was held and a committee established. In 1890, the Central Committee was formed by representatives from New York City and Philadelphia. Beginning in March 1890, the Committee received $10,000 per month from de Hirsch to provide temporary assistance to immigrant Jews. The committee was later renamed the Baron de Hirsch Fund. The Fund articulated eight goals, among them to provide relief to needy immigrants, educate them in English and citizenship, train them in trades and vocations including agriculture, and improve housing conditions in the congested immigrant sections of New York City.[10]

The New York-based Baron de Hirsch Fund's first specifically agricultural efforts were directed toward revitalizing foundering Jewish farm colonies in New Jersey, established in the 1880s by the Russian socialist Am Olam movement. The Fund purchased a 5,300-acre tract in Woodbine to establish its own agro-industrial community. In 1894, the Baron de Hirsch Agricultural School was established in Woodbine, New Jersey. In its second year, a Connecticut student from Colchester was among the school's twenty-two pupils. Future Connecticut farmers would continue to number among its students.

In the United States, the Fund devoted substantial financial resources to the Jewish Agricultural and Industrial Aid Society (JAIAS), founded in 1900 in New York City and largely operated by American-born Jews of German descent.[11] Having achieved a measure of success in business, these German Jews worried about the high numbers of Eastern European Jews living in crowded, filthy conditions in a tiny area on the Lower East Side of Manhattan, and they feared that these impoverished Jews, with their foreign dress, language, diet, and customs, would arouse virulent anti-Semitism in America. Such sentiments helped drive the charitable impulse to move Jews out of the city to America's small towns and countryside. The JAIAS pursued a policy of dispersion, resettling Jewish immigrants from New York and other crowded cities to rural and suburban communities in which prospects for employment and assimilation prevailed. Under the auspices of the JAIAS, Jews were sent to become farmers in rural Connecticut, New York, and New Jersey and to far-flung places with picturesque names such as Bad Axe, Michigan or Painted Woods, North Dakota.[12]

(Top) Goldstein Barn in Pachaug, ca. 1950s;
Photos courtesy of Goldstein Family
(Bottom) Goldstein Farmhouse in Pachaug, 1943

As its name implied, the JAIAS originally adopted a two-pronged approach to resettlement that included promoting both farming and light industry. The organization's four stated objectives were:

(1) The encouragement and direction of agriculture among Jews in America and the removal of those working in crowded metropolitan sections to agricultural and industrial districts.

(2) The grant of loans to mechanics, artisans, and tradesmen to increase their earning capacity and to enable them to acquire homes in suburban agricultural and industrial districts.

(3) The removal of industries in congested districts by helping manufacturers and contractors transfer their industries outside the large cities.

(4) The encouragement of cooperative creameries and factories and enterprises for canning and preserving fruits and vegetables.[13]

By supporting both factory relocation and farming, the JAIAS looked to increase the economic vitality of rural communities, create a nearby market for perishable produce, and provide farmers with opportunities for additional employment during the winter months. Jewish farmers living near Colchester, New London, Norwich, Willimantic, and other areas where factories could be found commonly subsidized their farm income by working as tailors, shoemakers, or butchers, trades they had brought over from the old country. In 1891, *The New London Day* reported on the volume of garment industry piecework being done in the tiny farming community of Chesterfield, noting that Jewish locals could "be found at full blast" assembling such garments as suspenders, coats, and pants in their houses. The fact that many of the pieceworkers were recent immigrants did not escape the *Day* reporter, who compared the "plodding" ways of the "old settlers" to the newcomers' turning "their hands so readily to various industries."[14]

In Colchester, the JAIAS tried to assure the availability of such handiwork by purchasing a former rubber factory and converting it into a lace manufacturer. Negotiations with the factory's owner were long, and on the night before the scheduled real-estate closing a fire broke out, destroying the factory.[15] Such setbacks, along with the high costs associated with factory relocation, convinced the JAIAS to focus on the agricultural rather than the industrial aspect of its mission. In 1907, its organizers abandoned efforts to move industries out of urban areas and in 1922 renamed the association the Jewish Agricultural Society (JAS).

Ben Zion Dember, Colchester, 1944

The Jewish Farmer, May, 1912, published by the Jewish Agricultural and Industrial Aid Society, New York, New York

With the help of JAS loans, Jewish farmers fanned out across the state. Early on, the organization pinpointed two areas in which to establish independent Jewish-owned- and-run farms, one encompassing the towns of Colchester and Lebanon and the village of Chesterfield in Montville, and the other in the towns of Vernon, Ellington, and Somers. In each of these areas the Fund financed the construction of a synagogue, one in Chesterfield in 1892 and another in Ellington in 1915, knowing that such a facility was essential to the creation of a Jewish community. The "Back to the Land" movement could only succeed if Jewish farmers were able to worship, keep kosher, and maintain cultural ties in the Orthodox tradition.

In 1908, the JAS revamped its assistance program to include four innovative pilot programs that would prove essential building blocks in the success of Jewish farmers. The four included the introduction of the Extension Department (responsible for a network of paid, Yiddish-speaking Jewish farm agents), the Employment Bureau (which placed Jews on working farms where they could earn wages and gain agricultural experience), Agricultural Education Scholarships for American-born children of Jewish farmers, and the publication of *The Jewish Farmer* farming journal.[16] The JAS felt strongly that periodic visits to individual farmers by Yiddish-speaking agricultural experts and lectures on agricultural subjects would allow for instruction on best farming methods. Kenneth Libo, a historian who grew up on a Jewish poultry farm in Lisbon, remembered regular visits from the Yiddish-speaking agricultural agent who offered advice over *a glesele te* (a glass of tea).[17]

In addition to the direct assistance provided by the Jewish farm agent, the publication of a monthly farm magazine both in Yiddish *(Der Yiddishe Farmer)* and English provided farmers with basic information. Advertised as the first Yiddish farm magazine in the world, *The Jewish Farmer* often reprinted illustrations with English captions from other farming journals. Many common farm vocabulary words lacked Yiddish translations, so the English words were used. Originally printed just in Yiddish, the magazine gradually took on English text and eventually was printed solely in English. The magazine helped explain to non-English-speaking Jewish farmers very pragmatic, simple farming techniques such as how to stack hay or plow and it helped to relieve the loneliness of farm life by reporting on social events, fairs, and prize-winning crops, all in a familiar language. It also encouraged the English-speaking children of Jewish farmers to take advantage of all agricultural education programs offered by the state's agricultural college and extension service by offering scholarships, holding contests, and publicizing summer classes. In Connecticut, the agricultural college was established at the University of Connecticut in Storrs.

After 1908, the JAS more carefully regulated its mortgage programs by providing not only money but also farm agents who would help immigrants locate and purchase farms appropriate in size and type to their individual needs and skills. Farm agents also helped these

Chicken shelters, Rashell and Cantor Farm, Ellington, ca.1940s

new Jewish farmers obtain farms featuring houses and barns along with household furnishings and farming equipment. Farm wagons, cars, and trucks are often cited in the deeds for such properties. "It is more important to help a man locate on a farm on which he can make a living than to help him with a loan on a farm on which he cannot make a living," explained JAS General Manager Leonard G. Robinson.[18]

In granting mortgages, the JAS frequently wrote into the deeds stern stipulations the farmer was required to follow, such as the following directives:

> Keep the property insured against fire
> Occupy the premises, reside there
> Commit no waste thereon
> Engage in no change in ownership
> Cultivate said farm
> Pay the taxes [19]

In exchange, the JAS agreed to take a second or third position among the creditors. One deed gave a farmer $1,200 for 150 acres of land with buildings in Columbia and Hebron that already was encumbered with a first mortgage of $2,100 and a lien of $1,900. The schedule of payments of principal called for an amount of $150 annually from 1924 to 1927; $200 annually thereafter, and interest at five percent.[20] The JAS screened potential applicants and rejected mortgage applications from individuals who had less than $400 to $1,200 in capital, but some farmers still defaulted, leaving the JAS saddled with bad debts. A 1923 *Hartford Courant* article reported that the JAS sued Jacob Fine for $3,565.41 due on a mortgage on a farm in Thompsonville. The JAS won the suit but had already been forced to pay the taxes and other bills for more than a year.[21]

Chesterfield and the Creamery

Strung out for a mile along state Route 85 on the way to New London and the Connecticut seashore, Chesterfield, Connecticut, today offers the beachgoer a chance to stop for gas and snacks at the general store. Now not much more than a rural "four corners," in the late nineteenth and early twentieth centuries Chesterfield was home to Connecticut's earliest known cooperative Jewish farm business, the Chesterfield Creamery.[1]

The New York City office of the Central Committee of the Baron de Hirsch Fund helped a small group of Russian Jewish settlers; a religious congregation known as the Society Agudas Achim from Brooklyn, New York, build a synagogue and a creamery business in Chesterfield. The board of the Central Committee believed very strongly in cultivating businesses that would support Jews willing to resettle outside of New York City. Too expensive for a single farmer to construct, a creamery would, in the Central Committee's estimation, bolster the cash earnings of nearby Jewish farmers, take advantage of farmland that was suited to dairy farming, and attract additional Jewish settlers.[2]

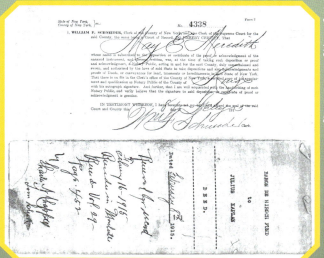

Chesterfield Creamery deed showing sale to Julius Kaplan in 1915

In March 1892, the New England Hebrew Farmers' Creamery Association (NEHFCA) was constituted to run a creamery and cheese factory and to buy and sell all kinds of farm produce. The NEHFCA was an offshoot of the New England Hebrew Farmers of the Emanuel Society (NEHFES), originally known as the Society Agudas Achim. A mortgage between the NEHFCA and the Central Committee recorded on the Montville land records on June 3, 1892 identifies the creamery property as including a small parcel of land adjoining the synagogue, a creamery building, and various pieces of equipment. The Central Committee provided a mortgage for $3,000 at five percent interest, with $2,000 from the Central Committee and $1,000 from Jacob Henry Schiff (1847-1920), a banker and financier in New York City and an original member of the Central Committee.[3]

Built at the corner of the Hartford Turnpike and Flanders Road, the creamery operated under a succession of paid superintendents. Small creameries were built in New England to help farmers get their milk processed into cream, butter, or cheese and to transport it by train to nearby urban markets. In the summer, forty-quart metal milk cans of cream were sent to New London by train and, in the winter, as butter to New York City. *The New London Day* reported that in the early twentieth century the Chesterfield Creamery consumed 1,300 quarts of milk a day at the height of the spring season. Farmers received "one-and-a-half and two cents a quart" when the unprocessed milk was delivered to the creamery.[4]

Probably built in 1893, the Chesterfield Creamery was a large, one-and-a-half story frame building that resembled a barn with a louvered cupola at the roofline. Equipment used to operate the creamery included a ten-horsepower boiler, a power milk separator, receiving vats, a milk cooler, churns, and the iron shafting, pulleys, hangers, and belting needed to run the steam-powered equipment. Recent archaeological studies have shown that a small icehouse was built to store ice cut from nearby Powers Pond. Ice could be stored to keep milk cool in the summer months.

But the creamery was not a success. Requiring a paid superintendent, the enterprise also incurred operating expenses such as fire insurance premiums and property taxes. By June 4, 1912, the Baron de Hirsch Fund foreclosed on the property and took title through court proceedings.[5] Many circumstances may have contributed to the creamery's failure, including the rapid turnover of Jewish farmers in the area, the new dairy farmers' inexperience, and the unpredictability of the milk supply. As boarders arrived to spend summers in the country, Jewish farmers with lodgers kept milk for their own use and to sell to the visitors at higher prices than paid by the creamery, undermining the creamery's efficiency. *The New London Day* reported that the creamery's daily intake of milk dropped from 1,300 quarts to 200-300 quarts during the summer.[6] By the 1920s, small creameries also faced stiff competition from larger regional dairies that sold not only cream and butter but also bottled milk, which was becoming popular for children because of its nutritional value.

Instead of being sold at public auction as originally planned by the Baron de Hirsch Fund administrators, the creamery building was sold to Julius Kaplan in 1915 for $300 in cash plus $25 in expenses.[7] Mr. Kaplan acted on behalf of a group of Jewish farmers that had organized as the Independent Hebrew Farmers Jewish Association of Chesterfield.[8] He ultimately sold the building to Abraham Miller, a butcher who had come to America from Russia in 1904.[9] Mr. Miller remodeled the building into a residence for his family by adding a full second floor and open porch. His daughter, Rebecca Miller Galper ran the homestead as Galper's Inn until it burned in the mid-1940s. The ruins of the creamery remain and are part of the State Archaeological Preserve that also protects the adjacent synagogue parcel.

Auction flyer, 1915

---

[1] An article in *The New London Day*, 1936, quotes an earlier article from 1890 that names nine individuals who moved from New York City to Chesterfield in 1890. Family tradition maintains that Harris (Hirsch) Kaplan, a whiskey dealer from Pereyaslov, Ukraine, had immigrated to New York City in 1887 and led a small group known as the Society Agudas Achim (Community of Brethren) from Williamsburg, Brooklyn to Chesterfield in 1890. In a letter written in 1901, Arthur Reichow, agent for the Central Committee (later the Baron de Hirsch Fund), states that in 1890 about twenty Jewish families from Russia "have upon my advice settled with their own means on so-called abandoned farms, scattered around Chesterfield & Oakdale." Settlers also may have heard about the area from Eastern European Jewish immigrants who were resettled in Connecticut beginning in 1882.
[2] Samuel Joseph, *History of the Baron de Hirsch Fund: The Americanization of the Jewish Immigrant* (Philadelphia: Jewish Publication Society, 1935, reprinted 1978), 34.
[3] Ibid. Also see H.L. Sabsovich, General Agent, Central Office of the Baron de Hirsch Fund to E.S. Benjamin, Esq., New York City, 14 December 1914, Baron de Hirsch Fund Records, American Jewish Historical Society (AJHS), New York, New York.
[4] Barnard L. Colby, "Chesterfield Settled by Jewish Refugees from Russia in 1890," *The New London Day*, 3 July 1936, 9.
[5] Sabsovich to Benjamin, 14 December 1914. Additional correspondence over a five-year period documents the Baron de Hirsch Fund's efforts to foreclose on the building and to sell it at auction. Letters in the Baron de Hirsch Fund Records, AJHS.
[6] Colby, 9.
[7] Ledger and Minutes Book of the New England Hebrew Farmers of the Emanuel Society of Chesterfield, CT, 1892-1933. The minutes were translated from Yiddish to English by Mark Nowogrodzki for the NEHFES. The ledger with the minutes is in the possession of Nancy R. Savin, President, NEHFES, Riverdale, New York.
[8] Sale agreement between Julius Kaplan and The Independent Hebrews Farmers Jewish Association of Chesterfield, 15 February 1915, Baron de Hirsch Fund Records, AJHS.
[9] Bruce Clouette and Ross Harper, Archaeological Preserve Designation Report, New England Hebrew Farmers of the Emanuel Society Synagogue and Creamery Site, July 2007, from the Archaeological Preserve Designation Report collection of the Connecticut Commission on Culture & Tourism, Hartford, Connecticut.

Cooperatives and Credit Unions

All farmers lived close to the edge economically, but as recent immigrants Jewish farmers could be especially hard hit by any misfortune. An article in the Hartford Courant in 1900 told the sad story of Abram Goldstein, a Russian Jew and farmer whose house burned down. "By thrift and economy he managed to rise above adversity," the reporter lamented. "It is a hard blow to him to be cleaned out of all his furniture and clothing which it has taken him years to obtain."[22]

For protection against the vicissitudes of life, Connecticut Jewish farmers turned to traditions of collectivism and cooperation. One of the first of these efforts occurred in the Rockville section of the town of Vernon, where Jewish farmers established a mutual aid society in 1905. In 1907, the group was reconstituted as the Connecticut Jewish Farmers Association of Ellington and developed programs to assist ill members, cooperatively purchase fertilizer, sell Jewish farm products, and even provide short-term loans in cases of extraordinary need.[23]

The JAS also encouraged its farmers to create community associations. Established by the JAS in 1909, the Federation of Jewish Farmers of America served as an umbrella organization for Jewish farm associations. By 1916, sixteen Connecticut communities had formed their own chapters. Members enjoyed social and religious gatherings as well as lectures on agricultural science and marketing strategies. The Federation also sponsored a Jewish agricultural fair in New York, attended in its inaugural year by 225 farmers from seven states, including farmers from Ellington who won a gold medal for presenting the "best and largest" number of exhibits.[24]

Baby chicks under gas brooders at the Cohen farm, Columbia, ca. 1940s; Photo courtesy of Cohen Family

Photo courtesy of Geretenish Exhibit

Membership in the Federation allowed its chapters to form local credit unions. Upon application, the JAS loaned individual Federation chapters $1,000 at two-percent interest. Local organizers were responsible for raising an additional $500 through the sale of shares at $5 apiece. With this seed money, the credit union could begin operation, providing its members with six-month loans of up to $100 at an interest rate of six percent per year.[25] Farmers in Fairfield established the first JAS-supported credit union in 1911. That same year, one formed in Ellington. In 1912, two more formed in Colchester and Lebanon. By 1915, 19 JAS-sponsored credit unions operated in the Northeast.[26]

CONNECTICUT JEWISH FARMERS' ASSOCIATIONS IN 1916

Town	Members	Name
Chesterfield	32	Independent Hebrew Farmers' Association
Colchester	147	Colchester Jewish Farmers' Association
Cornwall Bridge	24	Cornwall Bridge Jewish Farmers' Assoc.
Ellington	40	Connecticut Jewish Farmers' Assoc. of Ellington
Hebron	25	Hebrew Farmers' Assoc. of Turnerville
Lebanon	24	Lebanon Jewish Farmers' Association
New Haven	25	New Haven Jewish Farmers' Association
North Canton	13	North Canton Jewish Farmers' Association
Norwich,	55	Norwich Jewish Farmers' Association
Oakdale	65	Raymond Hill Jewish Farmers' Association
Rocky Hill	12	Rocky Hill Jewish Farmers' Association
Stepney	66	Jewish Farmers' Assoc. of Fairfield County
Storrs	15	Storrs Jewish Farmers' Association
Vernon	15	Vernon Jewish Farmers' Association
Willimantic	40	Willimantic Jewish Farmers' Association
Yantic	55	Yantic Jewish Farmers' Association

Taken from the *Guide to the United States for the Jewish Immigrant, A Nearly Literal Translation of the Second Yiddish Edition,* John Foster Carr, published by the Immigrant Publication Society of New York City, 1916.

(Top) Factory-style chicken coops, Cohen farm, constructed ca. 1942.
(Bottom) Cohen farm outbuildings, ca. 1930s;
Photos courtesy of Cohen Family

Photos courtesy of Geretenish Exhibit

The JAS's campaign to establish credit unions was a pioneering effort. Very few credit unions existed in the United States in the 1910s, Massachusetts having become the first state to pass enabling legislation for credit unions in 1909, and none existed to support agricultural communities. The need for farmer-organized credit unions declined after the passage of the Federal Farm Loan Act of 1916, which established Federal Land Banks to provide mortgage credit to farmers and ranchers, but by then the JAS's efforts to reform rural credit had been held up as a national model.[27]

By the 1920s, Connecticut Jewish farmers had begun to work collectively with their gentile neighbors, promoting the formation of cooperative purchasing and marketing associations. Farm cooperatives had a meaningful impact on the organization of agriculture in Connecticut. In 1924, Hartford County did more cooperative business than any other county east of the Rocky Mountains. Co-ops allowed farmers to save money by purchasing supplies in bulk and to maximize profits by controlling a large share of the market.[28] The Central Connecticut Farmers Cooperative Association, established in 1938, was open to farmers of all faiths, but its founders were seventeen Jewish farmers.

Cousins Janice Kurs and Erwin Goldstein with a calf, ca. 1938; Photo courtesy of Goldstein Family

The group focused on lowering feed prices for poultry farmers by purchasing grains at reduced bulk prices and mixing them in accordance with the latest scientific advice, procured from the state's agricultural college at Storrs.[29] Jewish farmers were also among the membership of Connecticut's largest and most successful cooperative, the Connecticut Milk Producers Association, an organization whose extensive membership allowed it to control the supply of milk and to raise its price.[30]

The Connecticut Agricultural Landscape

The period of Jewish immigration coincided with intense demographic shifts in Connecticut as the small state was transformed into a center of industrial production. Between 1870 and 1900, Connecticut became a state of huge cotton mills, silk factories, hardware companies, a flourishing machine-tool industry, guns and weapons producers, and a number of small machine manufactories. In all, Connecticut boasted more than

Featured on a souvenir postcard, this abandoned colonial farmstead was a picturesque reminder of the state's declining farm communities, 1907

nine thousand manufacturers, all beckoning the state's farmers to leave their land for better-paying jobs in the cities.[31] That temptation, compounded by competition from larger, more productive farms in the western states, led to large-scale movement away from Connecticut's rural areas. According to officials at the agricultural college at Storrs, one million acres of farmland reverted back to woodland between 1870 and 1920.[32] The problem of so-called "abandoned" farms became so acute that the state's Board of Agriculture began to publish a list of farms for sale. The 1899 catalog included almost 150 farms covering nearly 14,000 acres spread across all eight counties.[33]

Even as those Yankee families packed up for the city, Jews and other immigrants from Italy, Poland, Scandinavia, and Canada moved in the opposite direction, from cities to the countryside. The rate of immigration to Connecticut's rural areas was so great that in 1928, every third farmer in Connecticut had been born in a foreign country. Only Minnesota and North Dakota had greater proportions of foreign-born farmers.[34]

While many of the newcomers came with farming experience, most had to adapt to the state's rocky fields and widespread poor soil fertility. Each immigrant group pursued a variety of agricultural endeavors, but Italians became especially well known as truck farmers who transported their perishable produce from orchards and vegetable gardens to the state's cities, while Polish farmers often raised tobacco and many Czechs and Finns pursued dairying.[35]

Tom Cohen with bull, Cohen farm, Columbia, ca. 1940s; Photo courtesy of the Cohen Family

The Jewish farmers who took to the land in the 1890s continued the subsistence farming practiced by their Yankee predecessors. They kept vegetable gardens, apple orchards, and a few animals, growing crops principally for their family's consumption and selling the surplus. But with the assistance of the JAS extension agents and *The Jewish Farmer*, they adopted a more market-driven approach that focused on intensive farming and specialized crop production made possible by new farming techniques devised through scientific experimentation.

Many Jewish settlers in the Ellington region prospered by specializing in tobacco, which thrived in the Connecticut River Valley's especially rich soil. Dairying proved another early success. In the late nineteenth century, farmers typically sold their milk to creameries that harvested the fat for butter production and disposed of the rest. But in 1911, the Agricultural Experiment Station in New Haven discovered Vitamin A in whole milk, and mothers were encouraged to include cow's milk in their children's diets. With the assistance of federal farm loans, Jewish farmers took advantage of this new market, purchasing purebred cows and boosting their milk production with scientifically determined feeding regiments. Like many Connecticut

Jacob Goldstein (right) with his brother on a pasture along the Goldstein farm, Pachaug, ca. 1930s; Photo courtesy of the Goldstein Family

farmers in the 1920s, Jewish farmers began large-scale poultry farming, a pursuit that required little land and that was encouraged by the University of Connecticut at Storrs, which displayed scientifically designed chicken coops in 1918. Jews did particularly well in the poultry industry, not only as farmers but also as feed suppliers and chicken dealers, carting chickens to wholesalers and retailers (many of whom also were Jews).

Sarah Rostow on farm in Bloomfield, 1913

Kalmon Bercowetz (left) bringing in cattle from Chicago, Illinois, to Bloomfield in the 1930s

Raising a Crop of Tourists

By the 1910s, many Jewish farmers had begun supplementing their income by providing lodging for Jewish summer guests. Farm families found themselves catering to fellow Jews who were drawn by the promise of fresh food prepared in kosher kitchens, local synagogues in which to worship, and a respite from smoldering, stifling summers in New York City tenements. The host farmers could be counted on to uphold Jewish dietary laws. They engaged the services of *schochets* (ritual slaughterers) for preparing fowl and meat, and by the 1920s kosher butcher shops could be found in small country communities such as Colchester. One former resident recalled, "There was a kosher butcher and I still remember where it was, a little house right across the street from the Colchester Laundry. The *schochets* were there to take care of the chickens. My father would bring them [the chickens] on the same bus my sister and I would take to school."[36]

The Broadway House, Colchester, ca. 1925

Photo courtesy of Mary Donohue

Located across from Lake Williams in Lebanon, the Grand Lake Lodge became the largest resort in town accommodating 250 guests. Postcard ca 1940s

In 1928 *Hartford Courant* reporter Isabel Foster described the summer influx of Jews to Colchester in central Connecticut, an area that quickly became known as the "Connecticut Catskills."

In the summer the population jumps to 10,000 because of the summer visitors from the tenements of New York. Those who do not like to see a dignified and quiet old town turned into a resort for city workers should soften their hearts by spending a hot week in August sleeping in a court room on the Upper East Side of New York. It will then be a matter of rejoicing to them that these poor people can get away to the green country for a few weeks. They will look with delight into the sheds labeled Banquet Hall where a long table and two benches serve as the only furniture and at the hammocks hung in the run-down apple orchard—even at the noisy groups who gather in the town's little stores.[37]

In Yiddish, a *koch-a-lein* ("cook for you") arrangement was typical. This meant that the guests cooked their own food in the farm kitchen or in a summer kitchen set up in another building (to keep the cooking from further heating living quarters). Summer kitchens were common among farm families who did home canning, and jelly and jam making and who cooked for extra hired men during the growing season. That setup allowed farmers to profit not only from collecting weekly rent but also by selling fresh ingredients to the guests. The farm family made room for extra guests any way they could, sometimes even moving out of their own bedrooms to sleep in the barn. Saul Mindel described the motley collection of buildings that made up the guesthouses at his parents' farm: "There were several bungalows. One was originally an icehouse. One was originally a chicken coop. The City of Norwich disposed of their trolley cars and we were able to obtain one and turn it into a bungalow."[38]

Farm boardinghouses that provided both meals and rooms were less common. One was Pincus Schwaitzberg's Elem Farm House in Mansfield. An advertisement for the farm calls it a "first class summer place [with an] elegant summer garden, fresh butter, milk, and eggs every day, and fine spring water."[39]

Vinnie Goldstein, wife of Benjamin Goldstein, Goldstein farm, Pachaug, ca. 1938; Photo courtesy of the Goldstein Family

Elem Farm House boarders traveled by train from Grand Central Station to Willimantic to be picked up by farm wagon or by the Chelsea boat to Norwich, then continued on by "car" (presumably streetcar) to their destination.[40] Unlike today's "fat farms," spas that promise guests will lose weight during their stay, Connecticut's early twentieth-century Jewish farms were known for putting weight on their city guests. This meant hard work for the farmer's wife. "My mother in the early days milked cows by hand. And you have to remember she cooked everything we ate. Nothing bought was ready made. She made farmer's cheese [that hung] on a line outside. She worked very hard," eastern Connecticut Jewish farmer Harvey Polinsky recalled.[41]

Photos courtesy of Mary Donohue

Banner Lodge thrived as a resort in Moodus from 1930s to the 1970s complete with the iconic swimming pool

From housing a few guests in their farmhouses, some places—more than thirty, according to Janice P. Cunningham and David F. Ransom in their 1998 research report *Back to the Land*—eventually developed into resorts. Resort buildings included boardinghouses, cottages, hotels, bungalows, and camps that often offered waterfront recreation, dining halls, and nightly entertainment. Famous Jewish entertainers such as Zero Mostel traveled a Jewish summer resort circuit that included venues in New York, Pennsylvania, and Connecticut. Mostel's first cousin Jack Banner built the famous Banner Lodge in 1934 on the site of his father's farm in Moodus; the resort was known for hosting prominent comedians.[42] Some of these family-oriented resorts survive and now serve as youth camps or religious retreat houses.

The swimming pool, Grand View Hotel, Moodus, ca. 1925. The Grand View may have begun as a boardinghouse under Jewish farmer Harry Greenberg but its heyday as a resort was under the next owners, Seymour and Phyllis Pivnick, in the 1920s

A Banner Lodge stage show in the large recreation hall, ca. 1950

Making a Success of It

In the decade that followed World War I, more than a thousand Jewish farm families settled in Connecticut. From 1900 to 1933, the JAS made 2,248 loans totaling $1,373,641 to 2,183 borrowers on 1,885 farms (this includes farmers who received more than one JAS loan).[43] Many more JAS loans were made after 1933. By taking to the land in an era of industrial expansion, these Jewish farmers bucked economic trends. Not surprisingly, many did not make it. Economic panic, depression, or collapse of farm prices occurred with regularity every decade and a half, compounding the problems new farmers already faced. Still, many not only survived but thrived.

What distinguished those Connecticut Jewish farmers who succeeded from those who failed? In some cases, it was just good luck: a good location with close proximity to urban areas (including Hartford, New Haven, New London, Norwich, and Willimantic) that provided demand for fresh eggs, milk, and meat and that afforded access to trolley, truck, and train to transport produce to market. The presence of rich soils, like those found in the Connecticut River Valley, allowed a select group of farmers in the Ellington region to cultivate profitable tobacco crops, while important nutritional discoveries such as milk's Vitamin A and the development of sanitary, scientifically designed farm buildings benefited all. The help of the Connecticut Agricultural College at Storrs, where the first Jewish student was enrolled in 1898, and the State Department of Agriculture and its farm agents ended the farmers' isolation and kept them informed. But Jewish farmers also had a powerful ally in the JAS, which provided its own farm agents, farm publications in both Yiddish and English, and, perhaps most importantly, money to lend, often as a second or third mortgage. These essentials all supported individual farm families' own ingenuity and business savvy in turning old Yankee farms into flourishing dairy or poultry farms or summer resorts.

But individual success and survival is only part of the Jewish farmers' story. In many cases, these hard-working and inventive individuals revived declining rural communities and transformed the public's stereotypical images of Jews. In a 1928 *Hartford Courant* article headlined "Jewish Farmers Prosper in Connecticut: Living Down Their Reputation as Mere Middlemen Who Do No Labor, They Have Shown That They Can Till the Soil and Make It Pay," reporter Isabel Foster credited the JAS and its sponsored farmers for their commercial success and for literally creating something out of nothing.[44] Indeed, Jewish farming and the work of the JAS represented more than the production of food crops. Their efforts were part of an astonishing international campaign to stem the immense hardship and deprivation suffered by Eastern European Jews. Using new, creative, untried programs, the JAS helped a desperate people find a new home, country, and life without sacrificing their religion. The JAS's efforts in helping Jews become farmers in America yielded remarkable results among Connecticut's Jewish farmers.

For Further Information

For general information on Eastern European Jewish immigration to the United States, the book *World of Our Fathers, The Journey of The East European Jews to America and the Life They Found and Made* (1976) by Irving Howe is still the place to start. The books *From Haven to Home: 350 Years of Jewish Life in America* (2004), edited by Michael W. Grunberger, and Hasia R. Diner's *A New Promised Land: A History of Jews in America* (2003) provide more recent analysis of the Jewish immigrant experience.

The legacy of Jewish farming in Connecticut is an important part of the state's agricultural history. Much of it was documented in 1998 in the pioneering research report *Back to the Land: Jewish Farms and Resorts in Connecticut, 1890-1945* by Janice P. Cunningham and David F. Ransom. But in the more than ten years since the report was published, many new sources for the study of Jewish farms in Connecticut have emerged. Most importantly, the American Jewish Historical Society, New York City, has inventoried and catalogued the records of the Baron de Hirsch Fund (incorporated in New York in 1891), and the Jewish Historical Society of Greater Hartford, West Hartford, has greatly expanded its collection of transcribed oral histories and photographs. The Lebanon Historical Society has oral histories and artifacts related to Jewish farmers, and the Colchester Historical Society has a collection of postcards of Jewish-owned farms and resorts. And in what borders on the miraculous for a historian, *The Hartford Courant* can be searched on line for the years 1764 to 1984. This allows researchers to digitally access articles searched by key word or phrase. Search terms such as "Jewish farmer" and "synagogue" generate a wealth of articles. Those who would better understand the difficulties of adapting to farm life, whether in Connecticut or a nearby state, should read the excellent *Jewish Farmers of the Catskills: A Century of Survival* (1995) by Abraham D. Lavender and Clarence B. Steinberg.

For more on the Connecticut experience, *I Remember Chesterfield: A Memoir* by Micki Savin describes one of Connecticut's earliest (1891) Jewish farming communities in detail. For a more scholarly treatment, see Richard Moss's essay "Jewish Farmers, Ethnic Identity, and Institutional Americanization in Turn-of-the-Century Connecticut," published in *Connecticut History*, Spring 2006 and Mark A. Raider's "Jewish Immigrant Farmers in the Connecticut River Valley: The Rockville Settlement," *American Jewish Archives*. The Connecticut State Library holds the records of the state's Department of Agriculture and maintains a page on Jewish farmers on its Web site (www.cslib.org).

The Anshei Israel Synagogue, a fully restored country synagogue that served the Lisbon Jewish farm community, is now part of the Lisbon Historical Society. Special arrangements may be made with the historical society to tour the building. The society has published the booklet *A Refuge in the Country: Anshei Israel Synagogue* (2006) by Erica Myers-Russo.

For a well-written, first-hand account of the Jewish summer resort business, Esterita Blumberg's book *Remember the Catskills: Tales by a Recovering Hotelkeeper* gives a vivid description.

The son of Jewish farmers, television producer Ken Simon maintains a Web site (www.simonpure.com/resort) featuring an extensive collection of images of Jewish resorts in Moodus and East Haddam and a blog on which visitors share memories.

For more detail on the international and national events surrounding the period and the Jewish farming movement, read the *History of the Baron de Hirsch Fund: The Americanization of the Jewish Immigrant* by Samuel Joseph (originally published in 1935 and reprinted in 1978) and *Our Jewish Farmers: The Story of the Jewish Agricultural Society* by Gabriel Davidson, published in 1943. Jack Glazier provides a detailed account of rural relocation efforts in *Dispersing the Ghetto: The Relocation of Jewish Immigrants Across America* (1998).

---

[1] Roger Daniels, *Coming to America: A History of Immigration and Ethnicity in American Life* (New York: Harper Perennial, 1990), 226.
[2] David G. Dalin and Jonathan Rosenbaum, *Making a Life, Building a Community: A History of the Jews of Hartford* (New York: Holmes & Meier, 1997), 48.
[3] For a vivid description of Jewish life in the shtetls of Eastern Europe and of the immense social and economic pressures that led to mass emigration by Jews to the United States, see Irving Howe with Kenneth Libo, *World of Our Fathers: The Journey of the East European Jews to America and the Life They Found and Made* (New York: Harcourt Brace Jovanovich, 1976). Recent research on the gender, age, and occupation of those Jews who left Europe for the United States can be found in Michael W. Grunberger, ed., *From Haven to Home: 350 Years of Jewish Life in America* (New York: George Braziller in association with the Library of Congress, 2004).

4 "Aiding Immigrants: The Society for the Aid of Russian Refugees," *Hartford Daily Courant*, 12 July 1882, 2.
5 Samuel Joseph, *History of the Baron de Hirsch Fund: The Americanization of the Jewish Immigrant* (Philadelphia: Jewish Publication Society, 1935, reprinted 1978), 3.
6 "The Russian Refugees," *Hartford Daily Courant*, 28 June 1882, 2.
7 "The Russian Refugees," *New York Times*, 5 June 1882, 2. For information about L.B. Haas, see Dalin and Rosenbaum, *Making a Life*, 24.
8 The best known of such relocation programs in the United States were the so-called "orphan trains," which transported more than 200,000 poor urban children, along with several thousand adults (mostly women), to work on mid-western farms between 1853 and the late 1920s. See Marilyn Irvin Holt, *The Orphan Trains: Placing Out in America* (Lincoln: University of Nebraska Press, 1992).
9 "Aiding Immigrants."
10 Joseph, *History of the Baron de Hirsch Fund*, 22.
11 Gabriel Davidson, *Our Jewish Farmers and The Story of the Jewish Agricultural Society* (New York: L.B. Fischer, 1943), 3-34.
12 For an overview of Jewish relocation programs in the United States and a discussion of the motivations behind them, see Jack Glazier, *Dispersing the Ghetto: The Relocation of Jewish Immigrants Across America* (Ithaca: Cornell University Press, 1998).
13 Joseph, *History of the Baron de Hirsch Fund*, 129.
14 Barnard L. Colby quoted in "Chesterfield Settled by Jewish Refugees from Russia in 1890," *The New London Day*, 3 July 1936, 8.
15 Records of the Baron de Hirsch Fund, Box 64: Connecticut-Colchester, 1892-1925, Collection of the American Jewish Historical Society, New York, New York.
16 Joseph, *History of the Baron de Hirsch Fund*, 138-139.
17 Kenneth Libo, "Recollections of a Connecticut Jewish Farmer's Son," *Midstream*, September/October 2007, 33.
18 "Solving the Problem of Making Farm Life Attractive: Jewish Agricultural and Industrial Aid Society Has Effectively Coped With the Lure of the City and Is Unable to Meet the Requests of People Who Wish to Go 'Back to the Land.'" *New York Times*, 5 May 1912, SM13.
19 Columbia Land Records, vol. 19, pp. 443-446, July 6, 1923, quoted in Cunningham and Ransom, *Back to the Land*, 74.
20 Ibid.
21 "Judgments Given for Foreclosures," *The Hartford Courant*, 20 October 1923, 6.
22 "Salem: The Hard Lot of a Struggling Farmer," *The Hartford Courant*, 21 December 1900, 12.
23 Mark A. Raider, "Jewish Immigrant Farmers in the Connecticut River Valley: The Rockville Settlement," *American Jewish Archives* 47, no. 2 (1995): 219-220.
24 Joseph, *History of the Baron de Hirsch Fund*, 142; "Ellington." *The Hartford Courant*, 1 December 1909, 17.
25 Minutes of the General Assembly of the Jewish Farmers' Cooperative Credit Union of Lebanon, 1912, 1913, 1914, Lebanon Historical Society.
26 Raider, "Jewish Immigrant Farmers in the Connecticut River Valley," 224.
27 For newspaper articles lauding the JAS effort to reform rural credit as a model program, see "Conference on Rural Progress: Interesting Meeting to be Held in Boston March 7," *The Hartford Courant*, 26 February 1913, 5; "Solving the Problem of Making Farm Life Attractive," *New York Times*, 5 May 1912, SM13. For a later example, see "Jewish Farmers Eliminate Loan Shark," *Hartford Courant*, 30 November 1919, 9. See also Raider, "Jewish Immigrant Farmers in the Connecticut River Valley," 224.
28 "Connecticut Farmer Cooperates: He Has Successfully Organized to Market Eggs, Milk, Tobacco, Vegetables and to Buy Such Farm Necessities As Feed and Fertilizer," *The Hartford Courant*, 22 May 1927, D3.
29 Cunningham and Ransom, *Back to the Land*, 29-30.
30 "Connecticut Farmer Cooperates."
31 Bruce Fraser, *The Land of Steady Habits* (Hartford: Connecticut Historical Commission, 1988), 41.
32 Isabel Foster, "Alien Invasion of Connecticut Farms: Every Third Farm in Connecticut Is Now in the Hands of a Man of Foreign Birth, Almost Rivaling Minnesota and North Dakota," *The Hartford Courant*, 15 January 1928, E3.
33 T.S. Gold, *Descriptive Catalogue: Farms in Connecticut for Sale* (Connecticut: Connecticut State Board of Agriculture, 1899).
34 Isabel Foster, "Jewish Farmers Prosper in Connecticut: Living Down Their Reputation as Mere Middlemen Who Do No Labor, They Have Shown That They Can Till the Soil and Make It Pay," *The Hartford Courant*, 5 February 1928, E1.
35 Ibid.
36 Marian Jaffe Major, interview by Kenneth Libo, May 2005, Jewish Farmers Oral History Collection, Jewish Historical Society of Greater Hartford (hereafter referred to as JHSGH), West Hartford, Connecticut.
37 Foster, "Alien Invasion of Connecticut Farms."
38 Saul Mindel, interview by Kenneth Libo, May 2005, Jewish Farmers Oral History Collection, JHSGH
39 Advertisement for Elem Farm House quoted in Cunningham and Ransom, *Back to the Land*, 31.
40 Ibid.
41 Harvey Polinsky, interview by Kenneth Libo, June 2005, Jewish Farmers Oral History Collection, JHSGH.
42 Cunningham and Ransom, *Back to the Land*, 97.
43 Joseph, *History of the Baron de Hirsch Fund*, 288.
44 Foster, "Jewish Farmers Prosper in Connecticut."

JEWISH FARMS AND RESORTS IN CONNECTICUT

1791 — Pale of Settlement is established in Russia by Catherine the Great

1840s — German Jews begin arriving in Connecticut

1843 — Public worship by Jews is allowed by Connecticut State Statutes. Congregations are formed in Hartford and New Haven

1881 — Alexander II, Czar of Russia, is assassinated

1881 — *Pogroms* (sanctioned violence against Jews) begin in the Pale and continue through 1907

1881 — Hebrew Emigrant Aid Society is founded in United States (later renamed Hebrew Immigrant Aid Society)

1882 — British National Mansion House Committee is formed in London to aid in international Jewish relief efforts and sends Jewish refugees to Hartford in 1882

1882 — Society for the Aid of Russian Refugees is formed in Hartford and first refugees arrive in Hartford

1890 — The Central Committee is formed in New York City at the request of the Baron de Hirsch, a German Jew and philanthropist. It later becomes the Baron de Hirsch Fund (BHF), made permanent in 1891 in New York City

1890 — A small Jewish farm community sponsored by the BHF begins in Chesterfield, Connecticut

1891 — Jewish Colonization Society is founded in London, England.

1892 — On May 6, the synagogue in Chesterfield is dedicated with the congregants incorporated as the New England Hebrew Farmers of the Emanuel Society. Funding for the synagogue is provided by the BHF

1894 — Baron de Hirsch Agricultural School is opened in Woodbine, New Jersey

1898 — First Jewish student enrolls at Connecticut Agricultural School (later the University of Connecticut) at Storrs

1900 — Jewish Agricultural and Industrial Aid Society (JAIAS) founded

1905 — Ellington, Vernon, Rockville, and Somers chosen as an area of settlement by the JAIAS. Rockville Jewish farmers form a mutual aid society

1908	*The Jewish Farmer* Yiddish farm magazine begins publication in New York City
1908	JAIAS begins using four farming pilot programs: Yiddish-speaking Jewish farm agents, the Yiddish farm magazine, a farm employment agency, and agricultural college scholarships
1909	Federation of Jewish Farmers of America is established by the JAIAS
1911	First Jewish farmers' credit union in the United States formed by JAIAS in Fairfield, Connecticut
1911	Agricultural Experiment Station in New Haven discoveres Vitamin A in whole milk, encouraging its inclusion in children's diets
1912	Knesseth Israel Synagogue is dedicated in Ellington. The BHF provides the funding for the construction of the synagogue
1915	About 1915, Jewish farm families begin to take in summer boarders as paying guests
1916	Federal Farm Loan Act is passed
1922	JAIAS is reorganized as Jewish Agricultural Society (JAS). Connecticut Federation of Jewish Farmers is organized
1922	Jewish resorts are first established in Connecticut. Resorts begin to include lodging, community halls, waterfront beaches and other activities such as swimming, cards, and dancing
1924	Restrictions on U.S. immigration quotas for Eastern European countries are imposed, severely limiting the number of Jewish immigrants
1929	Great Depression begins
1937	First Jewish refugees from Nazi Germany arrive in the United States
1938	Central Connecticut Farmer's Cooperative Association founded in New Haven
1945	Allied forces win World War II. Postwar wave of Jewish immigration to Connecticut begins
1959	*The Jewish Farmer* ends publication
2006	Anshei Israel Synagogue, Lisbon, Connecticut, is restored and opened as a museum under the care of the Lisbon Historical Society

TIME LINE

"My daughter would say, 'Ma, you're taking care of the chickens better than a nurse.' I used to say, 'Listen, that's our bread and butter.' We had mortgages to pay."

Sarah Lauter, Colchester

Remembering a Life of the Land:
An Oral History of Connecticut's Jewish Farmers
Edited by Briann G. Greenfield and Lucas Karmazinas

The oral histories collected here teach us what it was like to feed chickens on a cold winter morning, milk a cow by hand, and wash the family laundry without the benefit of running water. Rich in detail and nuanced in their descriptions, they represent a powerful connection to the past, one that originates in the lived experiences of Connecticut's Jewish farmers who raised their crops and built their communities in the first half of the twentieth century.

The material gathered here came from several series of interviews conducted with current and former farmers, many from the rural Connecticut towns of Colchester and Lebanon, which acquired significant Jewish communities by the 1910s. The earliest interviews were conducted by the Jewish Historical Society of Greater Hartford in the 1980s. Others come from the collections of the Lebanon Historical Society, which conducted oral histories in 1998 with the assistance of the historian Dr. John F. Sutherland. More recent interviews were added to the collections of the Jewish Historical Society of Greater Hartford in 2005 by project historian Dr. Kenneth Libo, himself the descendant of a Jewish farming family in Lisbon, Connecticut. A list of all interviewees, their information, and the dates of their interviews appears at the end of the book. Because some of the interviews took place at a later date, the interviewees may now reside in a different location.

While the focus of this study is the Jewish farmer, it will quickly become apparent that farmers were part of a larger rural community that included butchers, bakers, store-owners, and even factory workers. Integral to the history of Jewish farming in Connecticut, these individuals recollections are included here, helping us understand the complex story of immigration and agriculture in Connecticut.

In preparing these oral histories for publication, every effort was made to balance the integrity of the original interviews with the demands of clarity and continuity. The editors organized passages by content, removed interviewers' questions and interruptions, arranged information by topic, and deleted off-topic comments and repetition. Tapes and transcriptions of the source interviews may be reviewed at the Lebanon Historical Society and the Jewish Historical Society of Greater Hartford, as can additional interviews with former Jewish farmers whose stories could not be printed here due to space constraints. In order to preserve each speaker's individual character, grammatical errors were left unaltered and both colloquial speech and Yiddish terms preserved within the text. While making decisions about what passages to select from the original interviews, the editors recognized that these texts are memory documents, meaning that they are not records of the past as it happened, but are instead produced from the perspective of many intervening years. As such, they may include some factual errors, but they also contain opinions, judgments, and reasoned analysis as interviewees look back on the past and reflect upon the changes they have experienced. In this regard, many of the interviewees stress the difficulty associated with rural life. Interviewees discussed living without electricity, the physical labor of farm life, and the need for every family member to contribute. Many portray this work as the basis for contemporary prosperity, as they describe how farming families used profits to send their children to college and on to more lucrative jobs in a white-collar world. But not all emphasize progress. Interviewees also recall a world in which families worked together, communities created their own organizations, and individuals lived close to the land. In this sense, these oral histories do more than tell us what happened. They also encourage us to think about our own experiences and consider what we have gained and what we have lost with the passing of these vibrant agricultural communities.

A FRESH START: "I WAS BORN IN BROOKLYN…"

Jewish immigration to rural Connecticut communities began as early as the 1890s and continued largely unabated through the early 1930s as the development of established communities attracted new residents. Few of these immigrants arrived in the United States with the intention to farm. For most, their first destination was New York City, though others made stops in Hartford or New Haven before taking to the country. Farming offered a chance to escape the crowded conditions of the city and also presented an opportunity to return to an agrarian life left behind in Europe. While some individuals received financial assistance from organizations such as the Baron de Hirsch Society, others scraped, saved, or borrowed from relatives in order to come up with enough money to get started. The accounts below are often told from the perspective of second- and third-generation Jewish immigrants who repeated for posterity their families' stories.

Irving Sol Kiotic, Lebanon

My parents settled in Lebanon on a farm [in the] early 1900s. They came on the farm for a definite reason. They originated from Poland. No Jewish person or family was ever allowed to own land in Poland or Russia at that time, and when they came to the United States, they saw an opportunity that they could own what they would call their own. They borrowed a little money from relatives, from friends, and my father went looking. He spent a lot of days, a lot of weeks, looking in different towns in Connecticut, and when he came to Lebanon, he said, "That's for me!"

My parents were not married in Poland. They came as individuals in this country. My mother was very young. Her family came here in, maybe before—I think it was 1898. My father came here as a young fellow. He was about sixteen years old. The only experience he had in Poland was driving horses because his family was in the horse business. As far as farming goes, he had no experience at all.

They came to New York City. My paternal grandfather, my father's father, was in the United States three different times. He was a United States citizen, and the first time he was eligible to vote was at the time McKinley ran for President. And he brought my father over to this country. There was what they called real estate brokers floating around the cities looking for contacts. They were looking to make some easy money, commission. In those days, commission was small and the money was big. They had connections. One man would be in New York and one man would be in the country. The New York man would find the customers, and the country man would find the sellers, and at that time the only sellers were so-called Connecticut Yankees. They referred to themselves as "Swamp Yankees." I don't know why. And that was the connection.

As I said, he came to this country at age sixteen, and the first thing he did was he started working in a so-called "sweatshop." He met my mother and they got married, both young. My mother was eighteen. My father was twenty-one, and they both decided that they would like to live in the country even though at that time a lot of parts of New York was country. Because, as I said, the real estate broker that he met had an accomplice in Connecticut. They were kind of bashful. They said, "Oh, an organization's not going to lend me money. I don't have any money, and they probably will not lend me money to buy a farm or land because they might think it's a waste, wouldn't be able to pay back."

Abraham Greenberg Homestead, Farmington, ca. 1900s

They settled, the first place was Bridgeport, and at that time, they just—something was wrong. They had a lease on property, which was not fertile land and my father observed it right off. He didn't want to waste any time with that. He gave a one-hundred-dollar deposit; he figured he'd better lose the hundred dollars than to lose years of time. So they did not go back to New York, but they came looking in Connecticut, and Lebanon was the only town that they went looking and they saw that property. It appealed to my father, and he made connections, borrowed money here and there from relatives and what they call *landsmen*. *Landsmen* is people coming from the same town or county, and they bought that farm for I think it was thirty-two hundred dollars. That was a lot of money for my father in those days.

My father owned land on three sides of Lake Williams. He bought a bare farm. A neighbor owned it. The house was empty. The barn was empty and the fields were neglected. So he bought the property, borrowed money, and he didn't want to look for any more from the friends or relatives, so he left my mother and an older child on the farm and he went to New York to work to earn some money to buy some cattle to go into the milk business. At that time it was not milk business, it was cream business. There was no outlet for milk in the Lebanon area.

Irving Bercowetz, West Hartford

They lived on Bellevue Street in Hartford. Bellevue Street was a very nice residential section. In fact Bellevue Street was a nice residential section when I was a child. My father kept a cow. Right on Bellevue Street. And he taught my mother how to milk the cow. She made her own cheese and cream and butter. My mother was a strapping strong woman, very, very energetic. My parents moved to Bloomfield in 1909. My mother had definite ideas about moving into the country. She was very apprehensive about it. She didn't want to move out. She wanted to be next to her sister. She wanted to live in the city. My father had different ideas. He wanted to fulfill his age-old dream of owning a farm, owning land.

My mother had definite ideas what she wanted to do. She wanted to make sure if she was going to the country that she had a Jewish neighbor, that there was transportation to get into the city, and that there was a store. Well, she looked with my father at a place where the Connecticut School for the Blind is today off Blue Hills Avenue, Holcomb Street, and there was twenty-eight acres there. That was too far out in the country for her at that time. She wouldn't move out there. But she decided that she would move out to Goodman Street and Cottage Grove Station in Bloomfield because she had a Jewish neighbor who was an older woman who was quite maternal to her. Mrs. Reiner. And there was a train that went to Hartford. At that time there were seventeen trains a day that went by our house. They bought eighty acres. That included—they paid eight thousand dollars for it—that included eighty acres of land, a house that was built around 1875, a horse barn, good-sized cow barn, sheds, livestock, and tools. And my father sold the livestock and tools and used that for his down payment on his farm.

Bernard Goldberg, West Hartford

My mother's maiden name was Rose Luchnich. Her father came to this country on a Baron de Hirsch loan, and she remained in New York with a sister. About 1905 she came through Ellis Island. My mother was a little girl. My father came when he was about twenty-one at approximately the same time, but they didn't know each other. My father met her when she was vacationing at her father's and her brother-in-law's farm, in Colchester. She had been living with her sister in New York and working there. Her brother-in-law was Sam Kabatchnick, and her father was Harry Luchnich, they ran a farm together.

My father was quite the wage earner in the old country and so he helped his parents and his brothers. He was a butcher. He was a butcher and he was a Zionist. But first he was a Zionist and then he was a butcher. And he helped his family come to America and then he came, the reverse of usual. He came to Colchester, directly to Colchester from landing in New York. He didn't stay in New York very long.

Rubin Cohen, Colchester

Most of them came from Poland, Russia. I used to hear my folks telling stories.... They came from Vilna Gubernia, a county in Poland. This county where they came from was bigger than the state of Connecticut. They had hard times. That is why they came to America.

Rachel Himmelstein, Colchester

I was born in Russia, May 15, 1897, as far as I know. I have no memories of Russia. Some friend of my father had bought a farm in Connecticut, and he was urging my parents to come out to Connecticut and settle there—buy a farm and be people in Connecticut. He had farmed in Russia. He had some land and farmed. That was what he did. From what I gather I know that he must have had a cow or perhaps more than one. He had saved up the money from what he was doing when he came to New York and took out a mortgage, which everybody did at that time. I know there was an organization but I really don't remember. I think that he must have received help because he must have gotten a mortgage somewheres. I remember hearing that, and I think that they may have been helping. In those days I was really too young to appreciate what they were saying. I must have been five or six years old because I went to school there.

Photo courtesy of the Geretenish Exhibit

Goldstein farm, Pachaug, ca. 1920s; Photo courtesy of the Goldstein Family

Lester Agranovitch, Colchester

My grandparents came to Chesterfield because of the Baron de Hirsch Fund. But the land was not good for farming. It was tough. So my grandfather went a few miles up the road to Colchester and he started a general store there and raised his family. The store was called the Agranovitch Store. They had some hardware, a soda fountain, boots. It was really a total general store. And he got along quite well. It became the center of the town of Colchester.

Rubin Cohen, Colchester

I was born in Brooklyn, New York on March 23, 1911. I was one year old when I came to Colchester. The Baron de Hirsch set up quite a few farms and sent quite a few Jewish refugees who came to this country to this area and helped them buy little farms and go into the dairy business. It was quite a Jewish community at that time—back in 1912 on. It didn't become a poultry area until about twenty-five years later. It was mainly dairy. It was an opportunity to get out of the city and live in the country and to make a living.

There was an old Jewish family here by the name of Kabatzcnik—he was more or less a real-estate agent. He used to find these farms—he was working with these people in New York with the Baron de Hirsch Fund. My folks moved in with this farm and they lived at this farm about four years. They couldn't make a go of it, so he turned property over to these people named Skut. No money changed hands. They just turned over the mortgage because he couldn't afford to pay the interest on the mortgage.

David Adler, Colchester

We moved to Connecticut in 1927. One of my brothers was sick at the time, and they recommended that we go out to the country to live, so we moved up here. He was a machine operator. They had a coat factory in Colchester. It was Mike Levine. They were contractors. In other words, they would send up the cloth to Colchester and cut it here. They would assemble it and ship it back to New York. I went to work in New York with my father as a machine operator... The family lived here, but there wasn't enough work out here, so he stayed in New York. His brother was a manufacturer of ladies' garments, so we worked for him. I'd come home on weekends.

Milton Virshup, Somers, ca. 1940s

Photo courtesy of JHSGH

Arthur Nassau, Avon

My uncle Hyman Virshup, an immigrant from Russia, was a graduate of the University of Michigan Agricultural School. He later settled in Ellenville, New York where he started a farm before moving, with the help of the Baron de Hirsch Fund, to a hundred and sixty-five acres in Somers. He ordered a kit to build a house from Sears and Roebuck. They unloaded it off Billings Road, and he built it himself very close to the road. The farm consisted of the house, two long chicken coops, and a few out buildings. He raised chickens with the help of his wife Henrietta and children Milton and Bernie.

LIFE ON THE FARM: "MOTHER WORKED VERY HARD…"

Despite the bucolic nature of the Connecticut countryside, rural life was often difficult. Homes were oppressively hot in the summer and bitterly cold in winter. Food was largely limited to what could be grown or raised on one's property, supplemented only by occasional purchases at the local grocer. But life was not static. In the 1920s, many farms received electricity. Motor trucks and tractors also became more common, connecting farmers to local markets and lightening the physical labor of farm work. Because these new technologies made such an impact on daily life, they figure prominently in interviewees' memories.

Belle Bercowetz Cutler, West Hartford

I was born on the farm. My mother had the assistance of a midwife. The house had seven rooms. There was no central heating, no running water, no electricity. The downstairs was heated by a large kitchen stove and another space heater in the dining room, which were both fueled with wood. Upstairs we used portable kerosene heaters. About 1912 a pipe-less furnace was installed in the dining room, which was in the center of the house. This too was fueled with wood. Now, perhaps you don't know what a pipe-less furnace was, but it was a furnace that had a grate on top that was about three feet by three feet, no pipes, and to get heat upstairs a hole was cut in the ceiling and grating was put on the floor up above and the heat went upstairs. Sometimes we still had to use the kerosene heaters.

We had no running water. Our only source in our home was a pump that we had to prime every time we wanted water. For bathing about once a week my mother brought in the huge galvanized bathtub for the adults and the small one for the children and water was heated on the stove and we bathed in the tub. About 1917 we did get running water. We had it powered by a gasoline motor and we had a toilet installed.

We always had cows and horses as far back as I can remember, and all of the children knew how to milk the cows and drive the horses and work with the horses. We learned that at a very young age. Every Thursday evening, this was when we were very young, Pa went to Hartford to Windsor Street to purchase the food. He came home with fish, a bag of bread, or really it was a sack of bread to last for the week, and food staples. Dairy products and meat came from the farm. There was no way of keeping the dairy products, but we did have means of keeping the vegetables we grew. We canned some. Root vegetables were put under soil in the cellar. Actually, at first we didn't even have a furnace. It was pretty cold anyway. There were barrels of apples, a keg of cider, which had been made from apples which we had sent to the cider mill. The children were very helpful. It was just part of our life. We gathered bushels of hickory nuts. For wine, we gathered dandelion tops, blackberries, wild grapes, and elderberries. My mother did. We canned huckleberries and blueberries for the winter. One year for Thanksgiving my father had the *shochet* kill our only goose. It was just convenient. I guess we had only one goose. He looked at it and said, "I knew him, I can't eat him."

Marion Jaffe Major's family, Lebanon, ca. 1960s

We always had hired help. And in the summer there was a great deal of pressure to take best advantage of the weather. So in addition to our regular hired man that lived with the family, we brought in extra help. Harry Kleinman used to live with us during the summer to help. But it was very important to take advantage of good weather during haying season because hay had to be absolutely dry before storing. Otherwise, we ran the risk of spontaneous combustion.

Marion Jaffe Major, Lebanon

[Mother] didn't want company. She was too busy. Busy jobs with washing the clothes. We didn't have a washing machine. And there was an old shed out here, I could just see it in my mind's eye, there is a big metal wash basin, and she would be standing there and rubbing and rubbing and rubbing, and then she would put the clothes on the stove to cook, to make sure they would become clean. There was no running water in the house so she took the pails of clothes, we helped, to the brook to rinse. The water was crystal clear and there was a big flat rock on which she beat the clothes with a flat wooden mallet, called a *prynick*, to remove excess soap. This stream turned into the Yantic River and was an important part of our lives. We went fishing there for trout but were happy with pickerel or sunfish. The stream was also our big bathtub and a place to cool off in the summer.

It wasn't easy to dress five girls in our family. My mother bought large bags of flour for baking her *challah*, especially the twisted kind for Sabbath, but the bags were strong and tightly woven. They became hand towels, pillowcases, and clothes, slips, for us girls. The girls who got married and left home were very helpful. Lena, the next to oldest, sent us clothes for school, skirts and jackets, and my oldest sister in Ohio rescued the flour bags and turned them into pillowcases with handmade button holes, they were great pillows! She also sewed them together and made bed sheets. I still have some. The pillows were stuffed with feathers from our own ducks and geese. Eventually the birds would end up on the dinner table, but their feathers were carefully plucked to be used in the pillows.

Harvey Polinsky, Jewett City

On our farm the men used the outhouse. The plumbing inside was for the women. Two of my jobs as a little boy was to bring in the wood for the kitchen stove and to shovel the path to the outhouse when it stopped snowing. We used the outhouse for many years. Our clothes and boots became very dirty while working on the farm. The rule was not to come into the house unless we were clean.

Growing up on the farm I remember that my mother worked very hard. She helped my father milk the cows by hand besides doing all the housework. Each day she made three full meals, which always included baked bread such as *challah, babka,* or biscuits. All meals were made from scratch because there were no ready-made products. She was very good at making farmer cheese by hanging the milk on the clothesline in a cheesecloth bag.

Irving Sol Kiotic, Lebanon

Every family who had any small plot of land did not have to go hungry if they had a little ambition to work that little plot. It didn't have to be an acre. It didn't have to be a half acre even. Ambitious people should not have gone hungry in the Depression. There was no need for it. My mother was a very quick blueberry picker or huckleberry picker. Those are two kinds of berries. She was a very good cook and she used to can huckleberry, applesauce, peaches. She cold-packed the peaches, which I liked best of all. We had a lot of Bartlett pears. We had other kinds of pears. We had hundreds of jars in the cellar. My father made shelves. Hundreds of jars accumulated. She even made piccadilli or piccalilli, whatever you call it. Anything and everything.

We bought meat. We would take the chickens, I told you, to the *shochet* and we went, Colchester was the nearest place. There was no freezers in those days, so it was done more frequent than the gentile people were able to do it, and we bought meat. The meat was good in those days. The farmers had a lot of corn and they fed the cattle corn. My father also was a believer in buckwheat, and buckwheat and corn made the chickens produce eggs like you would never think. Better than now with all the chemicals they put into the feed and everything.

Marion Jaffe Major, Lebanon

They worked the old-fashioned way. They worked with the horse and wagon. They took hay in by the pitchfork. And then they had a place in the barn with a big opening, and they would grab the hay, and then they would have it attached to a pulley, and my father would walk the horse to pull the hay up. There was a little mechanism on top of the barn where it was tripped, and it was run into the barn, glide into the barn I should say. So we had that for a long, long time, we had loose hay. But all of a sudden things got a little bit more modern, and my father bought an old-fashioned tractor. A big chore in the winter time [for] my father and my brother was to get the ice. They would go to the lake when it was very cold, and they would cut the ice, put it on the sled, and bring it home. They had these special saws. They would put all the ice into the icehouse, and they would put sawdust all around to preserve it and keep it from melting. So that's where all the ladies put all their goodies up in the ice house if there wasn't room in the iceboxes.

Irving Sol Kiotic, Lebanon

We had an icehouse, and we packed the ice so well that the ice lasted over the summer, and we had an icebox. The icebox didn't hold much food, but we would have [a kettle]—they called a washtub a kettle—right on top of the ice in the icehouse, and that's where my father would put some meat and chicken. We couldn't keep it long. A week the most, even with the ice.

Marion Jaffe Major's uncle, Lebanon, ca. 1930s-1940s

Photo courtesy of Marion Major

We lived near the pond, Lake Williams, and we always had an icehouse filled with ice in winter and we used that ice in summer. There was a so-called hook—you had to wait—we waited until the ice was about eight inches thick. If it was thicker than that, it was hard to handle. The size, the dimensions marked off the cut was eighteen by twenty-two inches. Now, we had like a hook my father made. He had a board about twenty feet long and two short boards, one on each end, which was twenty-two inches long. That was one measurement, and he would have the hook and put it and kind of stand on it, kind of scratch the ice to leave a mark. Then he had another board with two sides eighteen inches long for the second side. He would mark it the other way, and there was a saw, a handsaw. First you had to open the ice, open the pond near the mark with an axe, and then start cutting with a saw. It was about four or five feet long with handles where you could use two hands on it, one hand on each side, and saw. On a cold day when you sawed and water came up, your pants and overalls got like a board. They would stand by themselves because it had to be done in cold weather when the pond was froze. Then they would cut it, and then they had to use the same hook to pull it out. I have seen people pulling out ice and they slipped and they went into the pond. It was sad, but they were comical. You had nothing else to laugh about. Everything else was hard going.

Then they'd drive up with the horses—there was no trucks or anything in those days—and load up the wagon. First you had to have stakes on the sides of the wagon with a board so the ice would remain in the wagon. People who did their icing would have two people, one on each side of the cake of ice with a pair of tongs, and lift it up. My father always lifted it up by hand all by himself, no gloves, nothing. Never. Never used gloves, and some of the people would run over to watch how he did it, but he did do it. He didn't brag about it. They just wanted to see.

Then we'd unload it. The first three layers was easy. You'd slide it off the wagon on the board with sides on it so the cakes of ice wouldn't fall off the side. Slide it into the icehouse, and they were laid out and left about eight inches on each side for sawdust. The sawdust was the preserver to keep it from melting in summertime. Then when you go to above the fourth layer you had to go uphill with the cakes of ice into the icehouse from the wagon. Everything was hard.

Marion Jaffe Major, Lebanon

One of the most important parts of our lives was the garden. We depended on the harvest to supply us with vegetables for the rest of the year. We had a big sand box where we stored carrots, beets, potatoes, parsnips, and such for the winter. It was in the cool earth cellar along with all the jars of canned goods such as tomatoes for soups or as spaghetti sauce and, the most important, blueberry jam and jelly!

My mother would go to the garden at daybreak before the heat of the summer days. Of course the garden was very well fertilized by cow manure. The horse and cultivator led by my brother Milton made the land ready for planting. Mother worked very hard weeding, and I remember one day I brought her some juice to drink for her late breakfast. She took the cup and spilled it out, saying, "Can't you see I have no time? I'm busy."

Blueberries grew wild in our fields. We had our receptacles such as tin cans or pots that we attached to string and hung it around our necks [so] our hands would be free to reach out for the berries. The woods were full of blueberry bushes, and we often strayed and could not find our way home. We could finally see the telephone wires in the distance that helped us find our way home. The berries were so delicious. I ate more than I picked, it seems, by the looks of my blue lips and teeth.

Rubin Cohen, Colchester

We always had a garden and a chicken coop. When I bought this place, there was just one hand pump over here, and there was no electricity. I was the one that finally got the electricity to come through this road. When I bought this place there were about thirteen little farms on this road, and they all made milk. You know, one can or a can and a half. They were all Polish. We were the only Jewish family on the whole road. I went to the power company…that was when the government was helping to pay for rural electrification. I went to the power company and asked them to give us electricity on this little road, and they said if I could get each property owner to sign and guarantee them five dollars per month for electricity that they would put it in. So I went to every farm on the road, and everybody agreed but the next-door farm thought five dollars per month was outrageous. They held it up for about a year. Finally I went to the power company and said I will pay fifty dollars towards his rent. I more or less shamed the farmer into finally signing, and that is how we finally got power. It had to be forty-eight or forty-nine years ago. And this was a dirt road. You should have seen it in the spring.

Marion Jaffe Major, Lebanon

Many years ago we didn't have electricity but we had all these lamps all over the place. Some of them were beautiful. I have some upstairs, really ornate things. And this man, Mr. Jonan, came here and he said Jaffe, I want to supply this area with electricity, but I cannot afford the amount of money that has to be put into this project. Would you be willing to help out? Would you be willing to give us three hundred dollars? So my father said yes. So that's how the light and power started originally because he approached all the farmers around, and by the time he got through he had enough money to have this facility. Of course it wasn't as modern as it is now. You couldn't have two things going at the same time. If you had a light, for instance, in the living room, you couldn't have too many lights anywhere else because they would dim out. But little by little, they had these transformers and all the other equipment that is necessary for a smooth operation, and it was just fine.

The telephones came in a long, long time ago. It was just a little black box. It was right on the wall… with a little receiver…. If you wanted to telephone, you had to turn the handle. And it had all the numbers, and there were party wires—in other words—three or four parties were on the same line, and it would ring your number. Mine was one long number and four short ones, that was us— 47814. But if we were nosey and wanted to know what our neighbors were talking about, we could pick up the receiver and hear what they were saying. The sound was not quite as clear, so we said come on neighbor, hang up. But it worked out fairly well. Of course they had all these operators at some place that was doing all the work. If you wanted a number they would have to plug it in for you. You remember that, don't you? I don't know exactly how it worked.

Salowitz Farmhouse, East Lyme, 2009; Photo courtesy of Salowitz Family

Photo courtesy of Geretenish Exhibit

A Farmer's Life:
"The Wife and the Kids, the Whole Family Had to Help…"

The twentieth century ushered in a period of intense change in agriculture as new knowledge about animal feeding, breeds, and disease control combined with the introduction of mechanized farm equipment to facilitate increasing large-scale production and crop specialization. In the early years, many Jewish farmers made their money selling cream to local processors who produced butter and cheese. Whole milk did not become a marketable product until the 1910s, and then only with the assistance of the Jewish Agricultural Society, which helped secure regional buyers. But by the 1930s, the dairy market was in decline, and farmers turned to poultry. While some sold whole chickens, or broilers, as they were called, eggs became the primary products, and farmers built large chicken coops using the most up-to-date management methods advocated by the state's agricultural extension agencies. The recollections assembled here reflect these changes and illustrate the different kinds of farming operations Jewish farmers created.

Rachel Himmelstein, Colchester

We had cows and chickens and a couple of horses, and my parents raised their own vegetables. We had fruits, vegetables, raised corn of course… planted corn. We had that for the animals and for the winter. Of course markets in those days were not what they are now. We sold some eggs once in a while. There was a milk company who took our milk and had it in big cans. We raised vegetables, potatoes, corn… of course we had our own milk cows. We sold eggs occasionally. But we did sell. It was over one hundred acres. I am quite sure. One farmhouse and there were a barn, cow barn, and chicken coop. We had an icehouse because we had our own ice for the summer.

Irving Sol Kiotic, Lebanon

It was strictly agriculture, dairy and poultry, poultry on a small scale. All poultry was small. There was no outlet for eggs, either, because you wouldn't sell them to another farmer, and the cities were far away. There was no way of getting it there, except once in a while a buyer came through to buy eggs, and you had to be careful that you took cash from him, not IOU because IOU he'd never come back. There was a creamery in Lebanon. They would come around twice a week. The farmer knew when. He would have the cream ready in, I think it was ten-gallon—no, ten-gallon would be too much—ten-quart cans.

I remember my mother carried me in her arms, and in those days the mothers were very busy farming. They did not carry the child much after the child started to walk. About 1920. In those days when a child started to crawl, that child started earning his or her living. I remember we must have had about three hundred chickens at the time when my father gave me what today they call a beach pail or sand pail of corn to feed the chickens. I got into the chicken coop and they surrounded me. I couldn't move. I just stayed there until they cleaned out that pail of corn. That was chores.

We never had more than two or three calves at a time growing up. It was my duty, my job, to feed the calves, feed the chickens, and change the bedding under the calves. He could have had about fifteen or sixteen head of cattle and about twelve milkers. All the cows don't milk at once. I didn't milk till I was eleven years old. In those days it was done the hard way, squeeze by hand. Sometimes there was a so-called hard milker, you had to squeeze awful hard, and those were the cows I stayed away from. I told my father, that's your cow.

Cows on Sadek farm, Lisbon, ca. 1970s

You milk her. And it was in the bucket, and sometimes the flies would bother a cow, and she would kick and kick over the bucket, lose all the milk, and that meant a lot. That was a big loss if it was eight, nine, ten quarts. Cows did not give as much milk in those days as they do today. By breeding process and all that they upgraded cows that produce more milk. In those days we just had a cheesecloth that we put over the can and kind of used clothespins so it shouldn't get into the can, and the milk was dumped over the cheesecloth called a strainer.

We had a tank made from concrete, and we cooled the milk in that tank, and then we'd skim it when it got cold, the cream. That's when the Jewish Agricultural Society came in handy. They made contact with a milk company called the Providence Dairy, and they came to the Leonard's Bridge Station every day with a tank, milk tank, hitched to the railroad. So that's when a lot of farmers really started to breathe a little better for money. That's when we got a lot more per quart for milk produced on the farm. We sold it as fluid milk, no more cream. And in those days, little by little, the creamery went out of business in Lebanon for that reason, for the milk-market reason.

Kurt Hopfer, Norwich

I operated a dairy farm by the book. We increased the milk production by proper feeding, proper management. We raised our own replacements in the young stock—that means through artificial insemination, it matched the semen from a particular bull to a cow that you had, and succeeding generations were all better producers than their predecessors. At one time we had one of the top herds in the county as far as milk production. The farm is a very intensive operation. You don't have any time for anything. It's seven days a week from five in the morning until eleven o'clock at night, and sometimes you are sitting up with a cow.

Irving Sol Kiotic, Lebanon

The best technology was that my father, he happened to know—I never asked him how—that early-cut hay would produce the most milk and a certain crop like clover. In those days I never heard the word "alfalfa." Clover was a good milk producer, so instead of just planting so-called timothy hay and red-top, my father would buy clover seeds and plant clover, which is still considered one of the best foliage—am I using the right word—whatever, for producing milk. Roughage, I should have said.

1926, thereabouts. It started to get a little easier. We also had a truck to move the hay from the fields, and there was the so-called sulky plow. You didn't have to run after the walking plow. Used a pair of horses with a sulky plow. You would sit and go up and back, two plows, one each way, not both the same way. You went up and down the furrows. It started to get a little more modern. A couple years after that, they came out with hay loaders. We bought a hay rake, so-called. Farming was a little easier. Not good, but a little easier.

Dairy farming had been upgraded along production lines. Cows were giving more milk, selling more milk, and we heard of alfalfa, which is a very good milk-producing crop. We planted that. We had very good results. In fact, Extension Service held meetings on our farm. Farmers would come together that were interested in alfalfa. They would ask us questions, my father and I. We were no experts, but we told them what we knew, that's all. And we were doing very good. We had milking machines, after electricity, 1936 or so. That's when we enlarged the dairy business. That's when we built the new barn, 1938. And milk coolers by electric. No more ice. We didn't have to fall into the pond in cold weather anymore.

A milk truck came around …. That's when they started to talk about the milk tank, bulk-milk tank. The bulk-milk tank works the same way as cans, except you have an opening where you dump the milk. All the milk is together, no more individual cans. It was a different kind of a tank truck that came around to get the milk. The farmer's milk tank was calibrated a certain way that the driver from the milk truck would stick some kind of a measurement into the tank, and that would tell him how many gallons was in the tank. The driver would hitch it up to a hose, and he would start up a little engine, which sucked, vacuumed the milk into the large tank. The tank on the truck held thousands of gallons. He would go through a whole route to load up.

In those days there was a state law. You could not add anything to the milk or take anything out of the milk, but then there was so much milk that they started classes. The company started classes, Class one, two, three, four. First class was how much they sold in quarts. Consumptive use. Then they would have a surplus for cream. Then they'd have another class for butter. Oh, they cheated on all that. There's no getting away from that. Then the one thing that helped us—I don't know if other farmers took advantage—they would dump the skim milk, some of it, into the sewer. So they allowed the farmers to get the skim milk free if they paid the cartage. That is, paid the trucking company for bringing the milk back. I'm talking about when milk was still in the cans, forty-quart cans. We took advantage of them. We had about seven hundred chickens, which was quite a bit in 1932, 1933, and we didn't give the chickens any water, just milk, and the egg production was great, and we raised a lot of calves on that skim milk. I think we paid forty cents a can for forty quart.

David Adler, Colchester

Most of the dairy farmers turned to broilers, producing broilers. That was even before the war because the Army was taking a lot of broilers. And then when that died down and most of the big people went down South, they were able to produce the broilers cheaper down in Georgia, then they went into the egg business. From the broilers they went into the egg business, and even that had died down. Today, I would say, there are about two people left who sell eggs in Colchester.

Irving Sol Kiotic, Lebanon

There was peddlers from Hartford coming around. That's when things started getting a little better in the cities, especially during the Ethiopia invasion. When Mussolini invaded Ethiopia [in 1935], United States ammunitions factories started working, and when Franco, the civil war, business started picking up in the United States. That's when the so-called peddlers from Hartford and New London started coming around to the farmers buying eggs. The farmer never knew what the market price was. So whatever he could get, he took. You couldn't keep them forever. It's perishable merchandise. Perishable foods.

Marion Jaffe Major, Lebanon

We had a big—for those days—chicken coop, and it was right across the road. I can see my mother in the winter wearing her high felt boots through the snow carrying laying mash and water for the chickens. They rewarded us with many eggs. We managed to sell the surplus. In those days there were metal egg crates that were handled with care and sent parcel post most anywhere. We had regular customers.

What did we do when our supply of chickens and eggs needed replacing? The chickens took care of that. There were breeding hens that sat on the eggs for three weeks until they hatched. When the baby chicks finally hatched, we had a big box behind the kitchen stove and kept them warm and fed till they could go outdoors. We had little chicken houses where they stayed till they were big enough to join the adult hens. There were weasels, animals that would prey on the chickens. We had to be sure the coop was closed.

Sarah Laufer, Colchester

You tried to help, the same like the husband, to work with the chickens. Used to be a time we'd take in baby chicks and they used to catch a cold, you used to have to go at night to give them medicine and the feeding by hand in the beginning. Later we got the automatic feeder. But in the beginning we didn't have the feeders. The water used to freeze, in the beginning, so you'd have to carry water out to give the chickens. My daughter would say, "Ma, you're taking care of the chickens better than a nurse." I used to say, "Listen, that's our bread and butter." We had mortgages to pay.

Irving Sol Kiotic, Lebanon

One of the worst things was with the chickens. They called it coccidiosis, which was a killer for chickens. They cut it short and they named it coxie. They told the people how to raise them in very small individual coops, like fifty chickens or so in each coop, and then space the coops out over the fields, so the contact was minimal. But the disease was catching. If they walked into a coop with the manure from a chicken, then walked in another one, they carried the bug. It was a live thing. It made holes in the chickens intestines, and

Woman feeding chickens, ca. 1930s

Photo courtesy of JHSGH

Sheep on Sadek farm, Lisbon, ca. 1970s

they would bleed to death. So my father said, well, he'll try something else, keep them all in one chicken coop, which he had under an old barn there. The north side was sealed from the earth like a cellar on one side, and at that time they had a disinfectant called Creole [creosol] or something. It turned the water white. Now, whenever he went into that coop, he had a bucket there, he would step into it, the bucket. The liquid was about two or three inches deep. He would step into that and that would kill it, if there ever was any germs and bugs.

Jacob Laufer, Colchester

The wife and the kids, the whole family had to help to care for four to five thousand, three thousand chickens. He [the farmer] used to collect eggs, wash them, and spray them. They used to have a little grader by hand to grade eggs, you had to grade them separate. You have to candle them, you have to look through. With a candle. You look for blood, stones, cracks, sex. Every single egg, you have to weight them, and to know that this is a small one, a medium, a large, extra large.

Now you have everything automatic, they won't even weigh. That's why you couldn't keep more than a couple thousand chickens. And that's why you can't compete today. I couldn't stay in the business compared to today. Today you need everything automatic and everything is—before, the woman used to take two to three thousand chickens. Now two men can take one hundred twenty thousand, a big coop.

Harold Liebman, Lebanon

The poultry that we ran was egg production, and what we marketed on an ongoing basis were the eggs. We sold them in a number of ways through wholesale distributors. We had retail accounts or store accounts and obviously trying to get the best return that we could, and the more that we'd sell directly to the stores or to consumers, we'd get a better price. But on the other hand, a wholesale outfit was also needed.

Birds that we would sell would only be laying hens that had lived their useful life, and they had to be recycled or sold, making way for a new young flock. So that was something that may have occurred once a year. You'd have to sell thousands of birds. You'd have to get a dealer, and in those days there were local dealers and local dressing plants. There was one in Willimantic, which is no more there, but you know, we sold through the local outlet, but certainly not by a bird basis.

At the end we had thirty thousand laying hens, and that was not on what you call a modern basis with cages. There were, you know, on the floor running free—what do they call that?—free-range birds now. So to make the change to a more modern operation, I certainly gave it a lot of thought, but at my age, I felt it was not the time to make that kind of change. It would have involved half a million or a million dollars going into hock. So for a younger person, fine, but I didn't think that I should burden myself to that extent.

BUSINESS VENTURES (OR MAKING ENDS MEET): "YOU SEE, HE IS DIVERSIFIED…"

Working a farm was no way to get rich. Often, it was barely a way to get by. Many farmers supplemented their agricultural income with a variety of ventures. Some worked multiple jobs, taking piecework into their homes or traveling between urban factories and country farms; others started their own businesses, working as push-cart owners, shop keepers, or butchers; and still others opened up their homes as seasonal lodgings, inviting both relatives and strangers to live with them for the summer as boarders. For all, staying solvent required hard work and creativity.

Rachel Himmelstein, Colchester

During the summers we used to have what they called summer boarders in those days, who came out for the summer. Women with their children generally came out, and they stayed for several months sometimes. We not only rented rooms, but we fed them also. It was like a hotel. Nowadays you would call it a hotel. There must have been about twenty at times. June, July, August, sometimes it went into September a little bit. It was the summer months.

Jewish folks from New York City, sometimes they would stay the whole summer, mostly women and their children. We had a brook on our land, on our farm not far from the house, and they would spend the time swimming or bathing in the brook or sitting under the trees. That was it. Our hotel was not what hotels have turned out to be. People now come out for a weekend or a week or two. In those days it was very inexpensive. The boarders ate in a larger dining room. We had a large dining room. My mother did the cooking. Later when we had larger groups coming out, she used to get a cook for the summer. But there was plenty of work for the family, just to keep the place going. My sisters used to do the cooking, too, and my mother did the cooking, but when there were more people, of course you had to do more cooking so we did have a cook. I had two sisters, but one married and left the farm. I had two brothers. We had to feed the chickens. Wash things, utensils from the milk and so forth. We did a number of things. We were not idle. The boys were older and could do farm work. Feeding the chickens, sweeping the house… we did whatever we could, whatever we were able to do. We all chipped in.

Marion Jaffe Major, Lebanon

It seems that in the old days when we moved here there was not enough money coming in from the milk to pay the taxes, to pay the insurance, because these bills come when you own property, so my family started taking boarders in, too. There were two bungalows here that housed about four families. We had summer kitchens, still there, where they cooked. We spent our time as kids sleeping in the barn because our rooms started to be rented. And so the whole house was rented, and we did fairly well. And the most important part was we were poor but we were very happy. The living room had the old Victrola and the old plastic records, and we spent our time dancing. Every day we were dancing. After we stopped work the Victrola was going, and the young people would be dancing. Wasn't that something!

Irving Sol Kiotic, Lebanon

My family had—we had a lot of rooms. My family had a big income two ways. We had those who would rent the room, kitchen privileges, and then we had those my mother fed, room and board, and we had a big income from that, a real big income. I think at the end of the summer my mother gave my father a pouch, eighteen hundred dollars, and that was a bushel of money. That was from one season. Early twenties.

We had a Victrola way back in those days. Not many records for the Victrola. They were free to use that, and they had the whole farm to roam around on. There was a swimming hole up at Lake Williams, and my father had twenty boats, because we lived right on the lake, and he rented boats, fishermen, someone who was not living with us, for a dollar a day. The boarders had the privilege free to go boating, and there was islands. Beautiful islands and sandy beaches they can stroll on. They were swimming and even fishing if there was those who brought their own food, those who just rented a room.

They came as families, different sizes. Some women came [with] two children, some [with] three, and I think the most [of the children] that I remember [were] young because the older ones would stay in the city to work. Now the men would work all week and they would come to Leonard's Bridge Station by train from New York, and my father or someone else would go meet them with a horse and buggy.

And my father would do all the shopping for them. They all couldn't go to Colchester. The wagon wasn't big enough. Each one would give my father a list, and he never got paid for that. He never made his own list and charged them.

We had what you called a summer kitchen, which was not attached to the house. It was a cool building, and they had kitchen privilege to cook there, to eat there. They could eat outside, but they had the privilege of using it. *Koch-a-lein*, they cooked themselves. *A-lein* means yourself, *koch* cook. *Koch-a-lein*, cook yourself. *Koch-a-lein*. Until we sold the farm in 1924. That was the end. My mother said she had enough of that. The cooking. The *koch-a-leins* didn't bother her one bit, and the boarders didn't either, but it was work. She was a farm woman. She had to help my father with some things. [The boarders] would come two or three days after school was out and go back two or three days, four days before school started to get the kids clothes and whatever. Eighteen dollars a week for room and board. That was very reasonable. My mother was a very good cook. They were never able to eat everything she gave them. The cats and dogs lived a good life.

Rubin Cohen, Colchester

My folks, after they lost the farm they bought two buildings, and they keep summer boarders. That is what a lot of the farmers used to do here, they used to keep summer boarders. They would come out from New York. In those days I think they paid about fifteen dollars a week for room and board and everything. My mother used to work until three o'clock in the morning. Us kids used to sleep in the barn. What beds there were we had to give to the people.

I remember they used to come in by the train, but the train would stop coming here. That stopped running thirty years ago. It used to come from Amston. It would go from Middletown through East Hampton, to Amston and then to Colchester. Now the railroad station is a package store.

Dember family's Elmwood Hotel, Colchester, 1924

Some of them would stay all summer, and some would stay for a month. Some would stay for two weeks. Most all the farmers would keep summer boarders. There was a pond right in back of our place, and there were three or four lakes in the area. People would entertain that way. My mother and my sister, and later they had a helper, and they had another woman who worked and made the beds.

We had a couple of people—we never knew it because they seemed to be decent people—who came out and boarded at our place and also some of the other boarding houses. They would stay a month or two, and afterwards we found out they were gangsters. One of them, we seen on the news, was murdered. They were on their own, but they had visitors who came. But we never knew they were gangsters.

Harold Liebman, Lebanon

At that time there was a dairy farm, poultry farm, and a summer hotel. It's across from Lake Williams. It's still being operated as a resort. It was a summer resort open seasonally from Memorial Day through Labor Day. It was a family vacation place. Most of the clientele, most of, you could say, the middle-class people that we catered to, lived in the city in apartments, and this was an opportunity for them and their families to come out into the country. It was a family-oriented place. To me or to my sisters, it was an exciting time. It was in a different slice of life. You know, the food was good. There was an orchestra, a band. We catered up to about two hundred people. There was the lake with boats and a beach that was part of the operation. There was an orchestra, young people. We had what they called a social director who would see to it that everyone was happy all the time. For the children, they ran a day camp. There was a lot of activity in the sense that it was sort of an odd, artificial existence for me during that period of time, as compared to the lifestyle during the winter, during off season. But it was interesting.

While some individuals opened their farms to the resort business, others created businesses to serve the needs of farmers. Several Jewish grain dealers supplied dairy and poultry farmers with feed and other assorted goods.

Lester Agranovitch, Colchester

Yantic Grain was formed in 1920 by three different grain stores in the Norwich area. One was run by the two Polsky brothers who were in Yantic, which is an outlying district of Norwich. Then there were the Solomon brothers in Greeneville. Simon Solomon was my father-in-law. And third were Gurdon's grandfather and father Charlie and Sam Slosburg in Norwich. I'm quite sure it was Dan Polsky who decided that rather than be competitors, why don't they form one company. So they got together and formed the Yantic Grain Company on Cove Street in Norwich. Later, Abe Levin's father Michael, who was a businessman, bought out the interest of Morris Solomon and came into the business. Abe Levin had gone on to college and graduated from Yale. He was a young man when he started a grocery division in Yantic, and from that grew Universal Food Stores.

Yantic established a number of branches in various towns—New London, Simsbury, Ellington, Danbury, and Moosup in Connecticut and East Greenwich and Westerly in Rhode Island. With the help of the Agricultural Department of the University of Connecticut at Storrs, for whom Yantic Grain endowed a ten-thousand-dollar-a-year chair, we developed our own Big Y feeds.

Gurdon Slosberg, Norwich

My grandfather Charlie Slosberg was in the hay and grain business himself around 1900. And he developed a good business. I have some of his ads that he put out and ran in the paper. My father Sam started to work there when he was in high school.

Yantic Grain sold hay and grain mostly to dairy farmers and to some chicken farmers. We're on the Central Vermont Railroad. The feed would be transported on the Canadian National through Canada and come down on the Central Vermont to us in Norwich. We were the largest customers of Central Vermont in the area. We also sold barbed wire and whatever else a farmer would use to run a farm.

Brothers Morris (left) and Simon Solomon, owners and managers of the Greeneville Grain Company, later part of The Yantic Grain and Products Co., ca. 1925.

Photo courtesy of Nancy R. Savin

At Storrs they developed a high-energy feed called the Connecticut broiler ration that became known all over. They found that by using corn meal as the basis of carbohydrates instead of oats and supplementing it with the right vitamins, amino acids, and proteins, they could grow the fastest, cheapest.

Many broiler raisers worked on, let's say, a penny a bird a week. Some feed companies paid them to take care of the birds up to around ten, twelve weeks. Unfortunately what happened was that eventually all these broiler raisers couldn't compete with Georgia or Arkansas, where they didn't have to spend as much on fuel or insulation. So we ended up with mostly egg farmers here like Sam Koffkof, who is very big.

The Yantic plant included a huge silo building put up so that we could bring in soybean meal or corn in bulk on railroad cars and store it. We had bulk-feed trucks, which would go into a farm and blow the feed into the farmers' bins so that nobody had to lift a hundred-pound burlap bag. A lot of the farmers' wives at one time used to make garments or blankets out of those bags.

Connecticut's Jewish farmers also found creative ways to market their goods to increase the family income. Some established roadside stands, while others trucked kosher meats and farm products to urban markets. Still others never worked the soil at all and made their living selling farm products as grocers and butchers.

Rubin Cohen, Colchester

One of the first stores on the main street was owned by a man named Agranovitch. He ran a general store selling shoes, boots, newspapers, soda, clothing, etc. Joseph Agranovitch. He was in business before I even came here. He was Jewish. The merchants in town were mostly Jewish at that time. There is another man by the name of Abe Lezints. He ran a dry-goods store when I came here. He had a movie picture house upstairs, Tip Top Hall. Then there was Al Broder. He was also Jewish. He started a lumber-and-grain business. He was in business for years and years. Then there was Pete Cutler, also a Jewish man who was in the grain business and ran a little grocery store besides. He was there for years—Pete Cutler and Sons. Then there was Mintz. He ran a tavern before Prohibition. There was another merchant named Sam Stern that ran a dry goods store. They were the oldest Jewish merchants. Oh yes, there was a woman, Sadie Boretz. She used to buy up farms, give out mortgage money, foreclose on farms. She was a real sharpie. She would loan farmers money, then they couldn't pay, and she would foreclose. She left a lot of money.

There was another merchant. He was Polish. His name was Tony Ruttka. He had a meat market on Main Street at the same time as the Agranovitchs. He lived on a farm also, but he ran a meat market in Colchester. There was a woman by the name of Mrs. Berman, and Mrs. Krupnick, and their children. They used to pluck the chickens for something like ten cents apiece, right there at a little slaughterhouse on Mill Street. There was quite a rivalry between them. They would start plucking the chickens while they were still warm. The people paid them that owned the chickens. People used to bring them there. There wasn't any slaughtering going around at the farms. All these people were religious. They had to be slaughtered by a rabbi. It had to be done in this one slaughterhouse.

There were three bakers, Jewish bakers, fifty, sixty years ago. Rosenfeld the name was. I knew the boy very well. I used to go with him and peddle bread all day. Freezing on a horse and wagon. We would peddle bread all the way from Lebanon and Franklin.

Bernard Goldberg, West Hartford

Go right back to 1900 when the roads from Colchester to Hartford were dirt, and Hartford, a paved city, didn't grow their own vegetables. They didn't have their own chickens and eggs, and they naturally didn't have their own meat unless it was brought in. Colchester farmers looked for every avenue of earning to earn a living and then some,

Marian Selvin's mother (in truck) with Sunday visitors, Rocky Hill, ca. 1915

Photo courtesy of JHSGH

because they were thinking ahead. They're gonna send their kids to college. So they'd start out, farmers would start out with truck gardening, we'll say, certain things. One farmer might grow mostly cabbage, another might have potatoes, somebody else might have tomatoes. Whatever it was, summer or fall, or whatever the product was. Then there were those who would fatten a calf because they were dairy farmers, and they would take it to the slaughterhouse, where it would be slaughtered and koshered, and maybe they would have two calves. And they would put their product, whatever it was, whether it was meat or apples or canned stuff that they would can themselves, in their market wagon or their buggy, and they would start out in a caravan at six or seven or eight o'clock at night, depending on the season of the year, arrive in Hartford twelve to fourteen hours later.

I think most of it was around—well, it could be almost anywhere in the city. But a lot of the business was around Windsor Street. The Windsor Street area, where the Jewish people were. And they would sell their wares, or they had a regular customer for their wares, and they would start back. Sometimes a person would do it only once a week. But there were some businessmen, it took them a day and a half to get there and back. And some of them were real kosher, so they'd only have *challah* and milk on the way, so they'd have to stay home for a couple days for nourishment and rest and for whatever.

And then there were the hardy, who would go three times a week. They were businessmen. They would buy product from the Colchester farmers and re-sell it in Hartford. In the wintertime, see, they'd go in a caravan, because the highwaymen would rob them. At the turn of the century. This type of business lasted until, I would say, World War I. And then when the paved highway was built to transport Springfield rifles to the New London docks for World War I soldiers, it was a different story. Now you had a paved highway, and there were a few automobiles. And so there sprung another, the fellows who used to buy from three or four farmers and sell their wares in Hartford could now buy from fourteen farmers and sell their wares in Hartford. They sold their wares in Hartford and they had that little profit. They had the extra change, as they would say, they'd have the extra change to send Moishele to college. And not only Moishele, but they sent their girls to school, too.

Rubin Cohen, Colchester

A lot of these people when they came here…of course, in Poland they lived on farms, more or less. It was a good farm but in those days milk was two cents a quart, and you just couldn't make a living on it. Today, the few dairy farmers that are left are diversified. We have a farmer who makes more money on farm produce then he does on milk. He has a roadside stand. You see, he is diversified.

I remember I used to drive when I was a kid, there was a farmer here by the name of Hyman Levine. He used to sell eggs to all the restaurants, and he would slaughter kosher chickens. I used to help him. He had an old, old Ford truck. We used to go to Hartford at three o'clock in the morning and come back at eleven o'clock at night. He would sell eggs and kosher chickens. His son is still living. He now works for the Consumer Protection. He is a kosher meat inspector.

Daniel Goldberg, they used to call him Zadel, he was mayor of the borough three or four different times. He had quite a large family. They had a meat market and slaughterhouse. They had these big wheels. They would put a rope on one of the cow's hind legs and lift him right up. Then the rabbi would…the rabbi had a very sharp knife. If he had to use more than one stroke to cut the throat, the cow was not kosher. Same with the chickens. We had a slaughterhouse for the chickens years ago. Just a shed. Years ago they used ice for refrigeration.

Bernard Goldberg, West Hartford

[Father] was in the meat business. When he came to this country, he first worked for [Mr.] Cutler as a butcher. Cutler had groceries, grain, meats, and what have you. Then he couldn't get along with Mr. Cutler, so he went to work for Broder. Same type of store, meats, groceries, and some grain. And meantime there was a very, very religious and wealthy Jewish man in town by the name of Mintz who took quite a liking to Pop for many reasons, because he worked so hard for the Jewish causes. And he offered him a store that he had, an empty store that he had, to run a butcher shop, and so Pop took him up on it and ran the butcher shop in this particular store. But he didn't stay there long.

Pop was more in slaughtering than he was in retailing and so with certain help he had a slaughterhouse built, which was put up according to the modern slaughterhouses of the day, and he slaughtered for the Hartford community and for Colchester, of course, which didn't need very much, and for Willimantic, for Norwich, and for New London. There was more kosher butchering going on in Colchester than anywhere else in the state of Connecticut by that time. He was in slaughtering all the time. And occasionally we would have to have two *shochetim* for certain types of shipments of kosher meats, and we'd have one come in from Hartford to accompany ours, our Colchester man, who was well respected all over. Nachum Levine.

He would spend some time in the retail market. We re-inherited this market about 1934. "We," the family, because everybody worked in it. All the kids worked in the store. The person who had bought it couldn't stand the business depression in 1931, [19]32, [19]33, so in 1934 we re-inherited the store. The man owed Pop so much money for meats and for what-have-you, and Pop kept saying, "Stay another year, stay another year, I'll take care of you, you know times'll change, it'll be okay." So we took over the store, and people came from as far as Willimantic, from Middletown, from all over, especially for the kosher meats.

Irving Bercowetz, West Hartford

My dad didn't like to stay in the kosher meat market much. He liked to get out into

Tom Cohen, a butcher, peddling meat, Columbia, ca. 1930s

Photo courtesy of JHSGH

the country and deal and *hondle* with the farmers with the livestock. And he did the slaughtering and prepared the meat and brought it into Hartford. My parents had a very good rapport with the community. The Yankee and native population were always very, very nice. After he bought the farm he set himself up as a cattle dealer and a dairy farmer and a slaughterer.

The Yankees, as I said, were very good to the Russian and Polish and Jewish cattle [dealers] by extending credit at certain times when they purchased livestock so they could have additional capital to conduct their business. Money was always short. Money was something that was very scarce. George Woodford, who was then a state senator and health officer of Bloomfield, many times went to the rescue of my father to allow him to retain his slaughtering permit—and at no personal gain to Mr. Woodford.

We slaughtered cattle for well over fifty years. When my father first started slaughtering cattle kosher, they would take the rabbi, the *shochet,* to the farm where they bought the cattle. They didn't have the kind of inspection you have today in the slaughterhouses. Incidentally, the German Jews who were the business people when my father came and were well off and were doing very well gave the Russian and Polish Jews a good market and an easy entry because they didn't know the language, they didn't know the customs. Their children were not coming into the business. They were at the end of their business life, and the Russian and Polish Jews were taking over. And this same thing happened here lately, where the German Jews took over in the thirties from the Russian and Polish Jews, so you can see where the cycle is.

My father in the early years would travel as far as Norfolk, Connecticut, which I think is about forty miles away. He would buy several yokes of oxen and some other cattle, tie them together, take two days to walk them to Hartford to dress them. And on the way there were certain farms that they would stop. It was like a hotel for the cattle and for my father. They would serve him a non-meat meal, dairy, and vegetables. They had respect for his kosherness. They were unloaded right across from our house on Cottage Grove Station. And then we'd drive them up to our slaughtering plant where it is located now.

Ruth Adler, Colchester

My father was in the grocery business. I think they came to Colchester in about 1906 or 1907. They worked in the grocery store up until the early thirties. My mother passed away in 1929, and my father continued in the business until about 1932, 1933. I went to work in a dress factory. I was brought up in a grocery store. When I finished high school my mother had passed away already, and I could not leave my father. I was supposed to go to business school. I stayed in Colchester and went to work in the dress factory.

I was say ten to twelve years and I would be able to serve someone ice cream or get someone potatoes or something. Whatever was necessary. My father, besides the grocery, in the summertime he had a pony. In fact, when he first came to Colchester he went out with a pushcart by hand and foot. He had a route in the morning, and he went around, came back to the center and refilled, and then went out in the afternoon on a different route, and as time progressed he had a pony, and he went out mornings from eight or nine o'clock in the morning to one o'clock and did one route and spent the afternoon on another route. These were all to the places that had summer boarders or roomers. He had ice cream and candy, fruit and toys, and different things like that, and they were waiting for him to come. At that time everyone had to walk into the center to get ice cream and things like that, and it was not that easy.

Going back a little further, on Halls Hill Road there was a canning factory for a short period of time. I remember it but not too much about it. At that time my father used to have the store open until ten or eleven o'clock at night. People shopped at their convenience. It was mainly dry groceries and whatever produce we did get was brought in. I think Willimantic had a produce deliverer. As far as bananas, grapes, vegetables…farmers had their own gardens. There was not much you could sell to the public. He used to go to New York and get a lot of things. But also from the wholesalers in Willimantic. He used to buy all his toys and Christmas decorations, he used to get a catalog, and things were shipped from New Haven. There used to be a candy salesman that would come into town, but ice cream we would get out of New London.

Merchants Row, Colchester, ca. 1890s-1900s

Mutual-aid organizations also helped farmers compete economically. Credit unions such as the Colchester Free Loan Association allowed farmers to borrow money, while farmers' cooperatives provided mechanisms to purchase grain and fertilizer at reduced costs, build local processing plants, and market farm products collectively.

Rubin Cohen, Colchester

I'll tell you, what they did have here was called the Colchester Free Loan Association. I used to donate money to that all of the time. The old Jewish people used to come collecting, and I used to give them twenty-five dollars, fifty dollars, or sometimes one hundred dollars. They would loan people money interest free. In fact my father used to borrow money from them during the winter when there was no work, and when there were no boarders. I don't think it is in existence any longer, but they ran it like a bank. They loaned out money to needy Jewish people for no interest. They would get stuck once in a while, but mostly somebody had to co-sign a note. This was for sixty or seventy years.

Bernard Goldberg, West Hartford

Shortly after the Depression with the non-Jewish farmers, they got together and they formed an association, the Farmers Cooperative, and built a dairy, a creamery of their own, and started to send their milk to their own dairy, a farmers' cooperative. That could have been started in about 1934 or [19]35, I think, I may be a little bit early with it, I'm not sure. But there were all denominations of farmers, the Russians, the Poles, there were Italians, they were all being taken. The dairies were taking advantage of all of them. The dairies were profiteering. Well, they were paying one cent for two cents' worth of merchandise, and they were taking four cents back for it after they sold it. Today you have controls. Well, the utilities could have done almost anything they wanted then. They can't now, they have to ask for permission to raise your rates, not that it makes much difference. Sometimes they get more than they ask for.

As much as Jewish farmers embraced the rural life, many remained dependent on industry and the availability of factory work to supplement their agricultural income. While large cities such as New York and Hartford offered seasonal work in factories or piece work to be completed in the home, opportunities also existed locally. In Colchester, farming families found work at Levine and Levine Coat Factory, Schwartz & Schwartz (S&S) Leather Company, and the Aaron Dress Factory, often called simply "Cohen's" for its owner Isaac Cohen. The existence of these industries made Colchester one of the favored resettlement destinations of the Jewish Agricultural Society, which recognized that the presence of local manufacturers would only increase farmers' chances for success.

Irving Sol Kiotic, Lebanon

My father left his wife, my mother, and child on the farm, and he went back to New York to work to get money to buy some cows to get started. He borrowed what he could to buy the property, and he ran out of relatives to borrow money. He worked in the men's clothing shop, where they manufactured the men's clothes. They had in those days what they called seasons. Now maybe factories have the work all year. That time they had a season, three months, like for winter clothes. Then they would knock off a month and start the spring clothes or summer clothes. It was a cycle. So he worked a season, something like three months, and he lived poor in New York. When he was working, he saved every buck that he could, so that he could get started in agriculture.

Examining leather at the Schwartz & Schwartz Leather Company Factory, Colchester. Founded in 1906 by Aaron Schwartz as Schwartz & Schwartz Leather in New York, specializing in leather-covered wire hair curlers, Schwartz & Schwartz began supplying its leather remnants for do-it-yourself kits for producing comb cases, wallets, and small purses. Relocated to Colchester, the company provided numerous employment opportunities for farmers and their families.

Front office at Schwartz & Schwartz Leather Company, Colchester, ca. 1910

Ruth Adler, Colchester

I started working at a dress factory about 1931, 1932… before the NRA [National Recovery Administration] came in.[1] I was just making twenty cents an hour. We were working fifty-four hours a week, nine or ten dollars a week. I kept my job until 1968 because there were a variety of different employees by that time. Of course, the NRA came in, and there was an improvement. But during the Depression we got along. Let me put it that way. In World War I there was a shortage of sugar, flour, different things like that, but during the Depression we had everything. We had the stamps.[2]

Photos courtesy of JHSGH

David Adler, Colchester

I did work part-time in Levine and Levine after school. I worked one summer at my grandfather's farm. In 1929 Michael Levine was not here. It was just Levine and Levine. At that time he was just a contractor. A contractor got the goods from New York, made up the garments, and shipped them back. Little by little, we also started manufacturing. We sold coats. Of course my father also made garments to order. To order, plus the fact that we bought, also.

During the Depression I was working in the factory and, of course, there were a lot of people working in the S&S, working at Cohen's. That was the two big factories in Colchester. Levine's was more or less part-time. You would work making the garments for about five months, and spring garments you would make two or three, and the rest of the time you were off. It was seasonal work. The leather factory worked most of the time. I made about a quarter an hour at that time.

Stephen Schwartz, Colchester

The leather factory rented part of a large three-story brick building, which was at the time the largest building in Colchester. It was owned by, I believe, the Elgart family. The S&S Leather Company probably found it slow going in the twenties and thirties. In the twenties we developed a product which is this leather hair curler, and it was a soft piece of leather surrounding a piece of wire and enabled a woman to curl or "bob," as they say, her hair, so that in the twenties a flapper who had bottle curls could make her curls with these kinds of things, and they were apparently very popular. The finished product was taken up to New York and sold there. Raw material was picked up, the leather, brought back here, and the factory here would cut the leather and sew it… in many cases the leather-cutting took place remotely on farms where men would cut the leather and have raw leather brought to them, cut it, sort of swap the raw and the cut leather at some of the remote farms of the area.

The company gained some popularity or gained some size and degree of reputation as it began to manufacture leather wallets during the thirties and early forties. However, the company suffered a terrific blow in 1938 when this large brick building that housed the factory burned to the ground. It was only through, fortunately, the loyalty of the people who were the employees at the time and, as a matter of fact, the customers, that the company was able to survive during the later thirties and early forties.

The women didn't sew curlers at their own homes. In some cases there was no electricity or there was no facilities for it, so they had a row of sewing machines internally here and the raw material for them was an oval piece of leather that had already been cut by someone else. In some cases it was cut here, in some cases it might have been cut by their husband out on a farm. One of the gentlemen that I referred to before, who is still with us, and he is seventy-seven, remembers cutting curlers on a tree stump at his farm in Salem, which was about five or six miles outside of Colchester, it is almost a part of Colchester. It happens that he has six brothers, and they were all part of the business. They all, in one way or another, helped the company by working at home. They either cut, or they packaged, or they sorted leather, or they drove a truck back and forth from the farm back in to the company.

What was known as "home work" in the early stages of industry in this century was a very big business because people either didn't have the transportation to come to work or the inclination. There wasn't any reason to come to… there wasn't anything like benefits, or Social Security, or anything else, and the people all had chores to do on the farm. There's no telling how many people might have been employed on a home-work basis.

Rubin Cohen, Colchester

My father, he worked in the garment district in New York, and you couldn't make a living on the farm completely because it was seasonal. He would work about six months in New York, and the family stayed here. He worked at different garment factories in New York. He was a presser. Suits and coats. And then he worked at a coat factory in Colchester, Levine and Levine. They are still in business here. He worked there years ago during the season. The clothing business was all seasonal.

There was Aaron Dress Company that was owned by a Mr. Isaac Cohen. There were two brothers, Isaac and Harry, and Harry left. It is still standing. It is right next to the town hall. A fellow just bought it; it was caving in. He just restored it and made office buildings out of it. It was a big long building that went right around the town hall. This dress factory here must have been started in about 1920 to 1925. They used to cart the dresses to New York. They were shipped by Schuster's Express. They were a big company. They just sold out. They had six hundred units on the road. They still own property in Colchester. That is how they started the express. The father had a couple of trucks, and he contracted to carry these dresses to New York for Aaron Dress Company. From that they built up quite a big trucking business. They just sold out two years ago.

A CONSTANT YET EVOLVING FAITH: "EVERYBODY WENT TO *SHUL*. WELL, ALMOST EVERYBODY…"

Jewish immigrants coming to rural Connecticut maintained a determined effort to preserve and protect their faith. Concentrated communities, such as those in Colchester and Lebanon, allowed for the construction of synagogues, Hebrew schools, and community centers, gathering places for both socialization and worship. But while Jewish farmers practiced their faith, religious life was changing. By the 1950s, the Jewish Conservative movement challenged old Orthodox practices.

Bernard Goldberg, West Hartford

It was a nice Jewish community. They had a very nice Jewish society in Colchester. They had a semblance of a synagogue in my father, Daniel Goldberg, and a Pincus Cutler and a Leon Broder, Hersh Cohen, old man Dember. I can't remember his first name. And there was Aaron Feiden. They built themselves a *shul* [synagogue]. This was the nucleus—Orthodox, absolutely. And more Jews came to Colchester, and the *shul* wasn't satisfactory. So they bought a plot and—no, wait a minute, I think that Harry Strick gave them a piece of land near his bakery. And he, too, had a finger in the pie. They then took a carpenter from out of the Jewish community to be master carpenter of the project, to build a new *shul*.

These people would every day pick up a few youths, a few young people who would either carry boards, carry bricks, mix cement, bang nails, or what have you, under the guidance of the master carpenter—Glass. Say by 1910 or 1911, and eventually they got a *shul* built. They had a bathhouse alongside because they had to have a *mikveh* and what have you. So we as kids used to refer to that as the bathhouse.

They eventually built a better *mikveh,* a bigger one, in Zion Hall, but that was later in the 1920s, and they had a center *bimah* [raised platform where Torah is read], and almost all the artifacts were handmade. They got a lot of stuff. They brought much in from New York, but the synagogue had an upstairs for the women, and it was very easy for them to see the goings-on downstairs, to watch and to participate in the services.

Pop was a Zionist, he was one, but he was a good one. They didn't have money. In that *shtetl* it meant eventually that there would be a Jewish homeland, there would be an *Eretz Yisrael.* They made Zionists, so when they got here they had a few more people who were interested, who were serious about Zionism. They got members by force almost. He would pick up a crony, and they'd go out to the farms and to the people in town and tell them, look, this is how important it is to the Jews to have our own homeland, and someday we will have it, and we'd like you to be a member of the Zionist organization, whatever it was. They'd come back with dollars waving. They had a central fund, I believe, somewhere. I think it went through Hartford.

Let's say by 1920 there had to be very close to two hundred Jewish families in Colchester, a very Jewish community. By that time we had a new *shochet,* who was a very learned man and an excellent speaker. We had somebody else for Talmud Torah. A Talmud Torah wasn't inaugurated until about 1922 or so. The Jews in town built a community hall, which opened about that time, and they had two classrooms for Talmud Torah.

We had a *shochet,* a young *shochet,* from Israel. I remember his first name only, Naphtali. And he stayed in town a couple years and took off. And after that a Nachum Levine, who stayed in town until the late thirties, when he died. This man was an excellent speaker. Anything and everything. He was a learned Jew.

My mother was very kosher because her father was a Chassid. Observed all the holidays and observed all the rules and regulations and all the laws especially. Every business in town was closed and the streets were empty. Even on *Shabbos,* all the stores were closed on *Shabbos.* But there weren't many others. There weren't many others, so Saturday morning there was nothing doing on the street. Everybody went to *shul.* Well, almost everybody.

Marion Jaffe Major, Lebanon
Shabbos was something that my father took care of. He didn't do any work, nor my brother. The only thing they did was milk the cows because that had to be done, but they didn't do any other work—any plowing or anything else.

Photo courtesy of JHSGH

Lebanon synagogue, 1946

Bernard Goldberg, West Hartford

On *Purim* [a holiday] Pop used to give—I'm referring probably to nine, ten years old, and right straight through until I was nineteen or twenty even, every *Purim* Pop would dress up, and he would go through the town and even to the farms, and he would take a guide with him. A person, another Jew interested in Zionism or the Jewish National Fund. But everybody knew who he was the minute they saw the blue box, and the minute they knew that he wasn't there for nothing. That was his life. Jewish National Fund.[3] He did something all year round. His whole life centered around Zionism and the Jewish causes.

I told you how important it was for him to build a *shul*. He even went to Lebanon and helped them build their *shul* and raise funds. He went to Columbia and helped them with Haskel Rosen to build their *shul*. It's hard to say where, I can't even tell you because we never knew sometimes until after the fact that he had had a finger in the pie. But how much he contributed, how much time or what, it's hard for me to tell, because by that time I was a teenager and I was helping in the business as well as going to school. We knew that he was more than supportive of any Jewish function. I can remember one year when three families had to live off the business and our accountant said, "How could he do it? The business made twenty-six hundred dollars, and your father gave twenty-five hundred dollars to Israel."

Irving Sol Kiotic, Lebanon

[The synagogue] started in 1902, the first religious community High Holidays of 1902. I know just in which home it was held, too. The family's name was Kierman, K-I-E-R-M-A-N, and they gave us two rooms. In those days the men and women did not sit together. So in one room we had the Torah, the scroll. That's where the men sat and prayed. Houses were built to rent rooms. So there was a hallway and rooms on both sides. So across the hall is where the women sat, and they could hear the service going on.

Everyone walked to service. No riding. Everyone walked. There was a family that walked, the Himmelstein family, they must have walked three miles or more both ways. It was at… service was in that home until the man sold the farm. There was a lot of changes in farms and farmers. Then we went to another home, and we stayed there as long as they lived there. Then the next home was the Glotzerman home. We stayed there until they sold the farm. Then it went to the Kerachsky home, also two rooms in every house, and we stayed there until we had our own synagogue.

Our first synagogue, which was ours—no private homes—was a school. It was school number fourteen. The town of Lebanon gave it to us under the terms as long as we use it, free. We had very talented people who could lead the services. Very talented, right up until the end up here. Always, they were always, all the Jewish farmers were very educated, Hebrew and Yiddish. Very, very well educated. We had no reason for a rabbi.

The first was in our house, and there was a lapse of no Hebrew school until we built this last synagogue, which is now a church. I gave the land for it. After we sold the farm, I attended Hebrew school in Colchester. Then over there was a rabbi. When I got to attend, my family paid. There was school a couple years after I graduated. Let's say [19]34. It started to grow, and it grew an awful lot after World War II, when the refugees or displaced people started to come to Lebanon. We're talking now about Lebanon now, not Colchester. All the other towns, too, but in Lebanon it grew.

People from Lebanon, different parts, they started riding already and not walking. It was more convenient, and there was a little more education, explanation on what was said while we prayed in Hebrew. Very much social, very much. They had planned affairs, parties, get-togethers, meetings of different kinds. Always planning how to better, how to better ourselves, that is the organization.

The new synagogue burned during the High Holiday season of [19]55, and there was a land owner up Route 16 who made a living from summer boarders—boarders that is—and the summer was over, so she allowed us to use her social hall for the High Holiday services. That's how we finished the year, and immediately the organization committees, we thought that this—my father and I, my mother was not alive at the time, we thought this was sort of a good center for the town's Jewish people because in this section was more highly populated with Jewish people than the other sections. There was one section up on [Route] 87. There were two Jewish families and another section, too. Here there were twelve, fourteen, fifteen families within a two-mile radius. So we decided, what the heck, an acre of land wouldn't hurt us any. We gave it gratis, no money, free to the organization.

We all kicked in. We all kicked in money. We didn't go begging. We didn't sell anything. I think we decided five hundred dollars a family, two hundred fifty dollars for an individual. This is when the Jewish Agricultural Society came into play. I think we borrowed two thousand dollars from them. It was a small amount, and then the Gilman family helped us out. They gave us a whole year electricity free, gratis.

Bernard Goldberg, West Hartford

Every kid was bar mitzvahed. There was no such thing as no. At the Talmud Torah we had a junior congregation, the Talmud Torah, by 1922. And these young men, Joe Dember and Sam Cutler, controlled a young congregation, a youth congregation in Zion Hall. And we had a Torah there. It's a different generation. The change is the generation. The so-called modern Jewish mother is a good mother, but I don't think that she had the same values, so exact values, that her grandmother had. Oh boy, I wish I could explain it. The grandmother was much more Jewish and wanted more Judaism in the house and in the family, whereas the modern Jewish girl will tolerate a little bit more. She'll water it down.

Stephen Schwartz, Colchester

The Zion Hall was for many years the *mikveh* for the religious Jewish women in the town. I am not sure, for instance during the thirties and forties, how much use it had. I do know that in the early fifties, there became an undercurrent of Conservative Judaism rising up among the younger people who were becoming part of the leadership of the synagogue, and the clash between Conservative and Orthodox became apparent when, in this building… they sort of turned it into a community center and Hebrew school, and the community center began to hold Friday-night services around 1952 in a Conservative basis where then men and women sat together. And a lot of older men, and probably older women too, boycotted and wouldn't go because they wouldn't sit next to their husband during services and they were using the Conservative book. From 1952 until the time the synagogue, the current synagogue, was built in 1960, it took eight years for the Conservative undercurrent to gain enough momentum for them to build a new synagogue. During this period from 1956 to [19]60 when the synagogue was being converted from Orthodoxy to Conservativism, a splinter group of people who refused to become Conservative and wanted to stay Orthodox fought with the majority, left the synagogue, held their services in a converted shoemaker's shop down here on Lebanon Avenue, which is down from the synagogue and the bakery, and eventually raised enough money to build their own synagogue, which is the one you see over here. It is interesting to note that the synagogues were so close to each other the land almost abuts…the back part of the land almost abuts each other, and yet they have virtually no contact with each other because the surviving people on both sides will still not sit down and talk about the possibility of even coming back together.

Irving Sol Kiotic, Lebanon

There was a very learned Jewish man lived above us, related to the people who ran Grand Lake Hotel for many years, before it became a whatchamacallit, spa. And he was a very learned man in Hebrew. My mother spoke to him about starting a Hebrew school at our house. We were kind of centralized from the Jewish neighbors. Well, that was the arrangement. He ordered books. He got books, and he started a Hebrew school class in our house, kids after school. The neighboring kids would go there, and my mother had goodies every single day. She was a good cook, a good baker, donuts, pies, cakes, and even the gentile children that passed our house would come in for the goodies every day, every school day. That lasted until we moved away from that farm. That's where I got the beginning of my Hebrew and Yiddish learning.

Farmers Milton and Mickey Virshup, Somers, late 1940s

Arthur Nassau, Avon

Every Sunday for years the Nassau relatives would come for roast chicken dinner prepared by my Aunt Henrietta Nassau Virshup out in the country. Uncle Hyman Virshup was not religious, I never heard anyone talk about a bar mitzvah for his boys In fact, he used to kill the chickens for Sunday dinner behind the barn. When his mother-in-law discovered this crime she never ate there again.

Life in the community, however, was not exclusively Jewish. Although many Jewish farmers lived in proximity to one another and often congregated for religious and social reasons, they still remained an ethnic minority. As such, secular life, ranging from local events to public schooling, played an important part in the lives of Jewish families.

Harvey Polinsky, Jewett City

My parents were involved for many years in the Grange. There were very few Jewish people in the Grange even though it really wasn't religious in any way. It was more of a fellowship experience. The programs included dances, plays, and educational programs that were presented by invited speakers. Many people lived on small farms, and the Grange provided a special social environment for the family as well as being a powerful lobby for the farmer. My parents enjoyed the fellowship and took part in many of the programs, except they did not go to the dinners because of their commitment to being kosher. My parents made every effort to be part of the community. If the local churches or clubs needed milk or eggs, my father donated generously.

Marion Jaffe Major, Lebanon

We went to a one-room schoolhouse. Well, in those days there were not the schools like there are now, consolidated with all the teachers and all the pupils—in those days you went to a one-room schoolhouse, and the teacher taught all eight grades. The schoolhouse that we went to, and it was so cold in those days, I could just see my sister and I walking through the snow. The little schoolhouse is the first house on the left as you turned the corner.

Reverting to the one-room schoolhouse, there was a big stove. A belly stove and a big pipe going to the top of the ceiling. I can just see it. The school-teacher would get there early, and she had a couple of kids that would come early, too. They were strong and able to help her. And they would have the stove going. But we had to sit in our coats early in the day because it was just kind of cold. We learned—we were so motivated—she had flash cards for different grades, and she would teach them to add and subtract and multiply. She would have a sign that said "5 and 3" and you would have to say what the answer is. It would be on the other side, probably. But we studied hard. We studied at home, and we would test each other, and I'll tell you, my sister Lena went to a one-room schoolhouse, and she could spell better than anyone who graduates high school nowadays.

The teachers were very, very, dedicated to us kids. If you were a little slow, she would keep you after school for a little while. We had Miss Scalinski, because the teachers had to live with a family because they didn't have cars, and they had to live with a family that would support them, and couldn't be too far from the school because she would have to walk to the school.

We did have the privilege of going to school in Colchester. We actually belonged to the town of Lebanon, but Lebanon had a bus that had to go up that country road. It was not paved, and it was hilly, and my parents were afraid that the trip was not safe. So they paid a hundred dollars for each of us to go to Bacon Academy. However, my sister and I did nothing. We were not that smart. All we did was study. The kitchen table was full of our books. We would work in the kitchen because it was too cold in the rest of the house. It was full of books so we could study, study, study. So we both got very good marks, and after a while Mr. Roolin, I think his name was, I can see him in my mind's eye, said, Jaffe, I don't want your money for the girls, they are so bright, and they give us so much pleasure having them in our school that you can come here for free. Now wasn't that nice?

When we went to school there, we thought it was gorgeous. It was old-fashioned. It had a privy at the end, but the rooms upstairs, there was the English teacher, and the French teacher, and the math teacher, and the social studies teacher. I can just see them in my mind's eye, but they were very, very diligent. And we were motivated. The kids in those days wanted to learn.

Harold Liebman, Lebanon

There was a one-room schoolhouse for my first four grades. First four years. There were eight grades in the schoolhouse. For the second four years, these were closed, and we moved to a so-called two-room schoolhouse for the remainder of the eight years. Goshen Hill. The year that I graduated elementary school was the year that the so-called new elementary school was opened in town. That was a consolidated building, and all of the regional schools were closed.

I think there were rows. The younger children were in one row and then you went up all to the eighth grade. I'm sure—I don't remember, but I'm sure there were several grades that were missed. There may not have been enough children. And there was a teacher, and there was an outhouse, of course. There was no running water. There was a stove, and I used to walk. This was a dirt road, of course.

END OF AN ERA:
"IT IS NOT A FARMING COMMUNITY LIKE IT USED TO BE."

For most of the oral history interviewees whose stories are collected here, farming is a way of life long since past. Looking back on the experience of farm life, interviewees considered the legacy of farming today.

Bernard Goldberg, West Hartford

The farmers had their problems. They were dairy farmers. These are the men of the Baron de Hirsch loans. What I think I know. Baron de Hirsch was a French Jew who was the railroad czar practically of Europe, at least, we'll say Germany, Poland, Italy, France, Spain. He set aside a phenomenal amount, six million dollars, at a time when people were working for three dollars a month back in 1880. And the idea behind it was to finance the oppressed Jews out of Poland and Russia, out of Russia and that part which is now Poland, to finance their way to a new life in America. Now, we had quotas in the States, and the only persons who could come into the country outside of the quotas would be farmers, and so he bought farms to settle these people on. But he didn't buy farms in Iowa, where you paid, we'll say fifty dollars an acre, or in Kansas where it was also fifty dollars an acre. He bought farms in Connecticut that were rocky as the devil at maybe two dollars an acre. This way he could bring in more Jews, but these farms were only suitable for dairy cattle. You couldn't raise and till or cut grain and what have you. So these farmers made their meager living from their dairy stock, who ate grass around the stones. And they'd clear a field and have a little corn for them for whenever, in the winter, or what have you.

They were getting by in 1930 and 1931. They were getting this tremendous [fifty-cent] piece for [a] forty-quart can of milk. A cent and a quarter for a quart of milk. Out of this money they saved even enough to send their children to college, and ninety percent of them amounted to not just something, became pillars of their communities, wherever they were, in New Haven, Hartford, New York, Baltimore, they went all over.

Rubin Cohen, Colchester

I must have thirty Jewish farmers that came here. The Nelber family. There were three or four brothers; Louie, Ike. There was a Berminitz on Route 354, Solomon and Gilman. Stolman on Old Hebron Road. His son still runs the golf course here in Colchester. He made a lot of money in the poultry business. Family name of Jaffe in Lebanon. Schaeffer on the Swager farm. Slovkin had a farm and a dry-good store for fifty years. Himmelstein brothers. Goldberg on Buckley Hill. This is going back fifty or sixty years. Shavitz on the West Road. Glovenger on Lebanon Road. Lerman on Buckley Road. Mintz brothers on the Parent Road. Marx on Chester Hill Road.

Most of them did come from the Baron and also Jewish organizations in New York. They all came out of New York City. You know what is amazing is how most of these farmers' children became doctors and lawyers, professional people. They went off to college. Most of them worked their way through college and became very successful, which they don't do today. Youngsters today, very few are working their way through college. It's their parents that have got to slave and sweat to keep them in college. It was a very closely-knit community. People got along pretty good. Of course there were petty jealousies, but you will find them in any nationality.

Rachel Himmelstein, Colchester

My brother, both brothers farmed. One took over the family farm, and one bought his own farm. My other brother took over from my parents and farmed on that. Then he bought the George Mills farm, which was just about ten or fifteen minutes walk from the old farm, and he sold the original farm. After my brother sold it, the people who bought the farm did not live here, and somehow or other the farm burned down. They sold land, and many, many houses have been built. It is not a farming community like it used to be. My older [brother], who moved away and bought his own farm, he had two sons and a daughter. The daughter grew up and became a teacher. She taught in Lebanon. The two sons, one stayed on the farm, and the other became a school teacher...went to college and so forth, and married. He stayed on the farm for a little while but moved out to Hartford. None of them are farming now. But my brother who bought another farm out in Lebanon, one of his sons is still on the farm. Two of his sons are still farming. One, who became a teacher in Hartford, comes out to the farm to help when he can. Dairy farm. But they raise their own corn and vegetables and so forth. And his son is not going to be a farmer.

---

[1] The National Recovery Administration was a New Deal program under the Roosevelt administration that helped workers by setting minimum wages and maximum weekly hours. The United States Supreme Court ruled the NRA unconstitutional in the court case of Schechter Poultry Corp. v. U.S., but many of its labor provisions were incorporated in the Wagner Act of 1935.

[2] Food stamps were initiated by the Roosevelt administration in 1939 to ease hunger and supplement farm income.

[3] The Jewish National Fund is a Zionist organization, originally founded in 1901 to purchase and develop land for Jewish settlement in Israel.

"I was interested in people not only as images, but also as human beings. In stories that they would tell me or interviews I had with them. It seemed to be it was an important part of what I was trying to communicate."

Jack Delano, as quoted in 'Far from Main Stree

FARMERS ON FILM:
Photographs from the Collections of the Farm Securities Agency and the Office of War Information

Briann G. Greenfield

In 1940, Jack Delano traveled to Connecticut. It was the first visit to the state for the young photographer. Born in Kiev, Russia and raised in Philadelphia, Delano was surprised by the New England that he found. There were of course the white church steeples, town greens, and old Yankees he had expected to see. But Delano was struck more by the region's cultural diversity than by its traditional charms.

On his Connecticut tour, he encountered Polish tobacco workers in Enfield, French Canadians employed in Montville's paper mills, and Finnish poultry farmers settled in Canterbury. But Delano was particularly impressed with Connecticut's Jewish farmers. Twenty-five years later, in an interview in the collections of the Smithsonian Institution's Archives of American Art, he would recall "Jewish farmers, who would get up at the crack of dawn to look after their cows in their pastures, and we would be invited into the house where this patriarch with his long beard, looking very biblical with his wife and his skull cap, this same old man would go out and look after his crops, and so on, and would go to the little synagogue in Colchester, living a completely Jewish life." Today, Delano's photographs, along with those of his colleague John Collier, provide a remarkable visual record of those Jewish farmers and their experiences.

Collier and Delano's images belong to a larger body of photographs commissioned by the Farm Security Administration (FSA). The FSA was primarily a relief agency created by the federal government to assist poor farmers during the Dust Bowl and the Great Depression with loans, government aid, and resettlement projects. But the agency also included a photographic division devoted to recording the agency's history and with it the history of the rural communities it served. With the coming of World War II, the FSA's photographic division was transferred to the Office of War Information. The photographs of Connecticut's Jewish farmers record their existence through both of these eras.

Ostensibly, the photos were intended to document the work of the FSA and to publicize the agency and, later, the war effort. Many of the images can be understood in this narrow light. There is, for instance, the photograph taken by Delano in 1940 of a Jewish poultry farmer's truck that was purchased with FSA funds. A separate image of the truck's owner Mr. Paul Klappersack shows a strong, young man looking every part the worthy recipient of public funds.

By the early 1940s, Jewish Americans had become an important subject matter for the FSA. In addition to Delano and Collier's Connecticut trips, the Farm Securities Agency and the Office of War Information funded photographic surveys of Jewish farmers in New Jersey and New York, and departing from its traditional focus on rural life, an extensive survey of Jewish life in New York City.

The very existence of Jewish families making their lives in the United States made for good public relations material. From the beginning, the FSA had been primarily concerned with documenting rural cultures, but as World War II approached, the agency had to justify its existence by photographing material that could be used for war propaganda.

But these photographs are much more than propaganda pieces. Many of the photos are remarkable for their aesthetic quality. Delano was trained at the Pennsylvania Academy of Fine Arts and approached his photographic work through the tradition of formal portraiture. While many documentary photographers sought candid moments in an effort to display their subjects in the rawest possible light, Delano produced deliberate and studied portraits of his subjects. His work shows a complex portrayal of Connecticut's Jewish community: Old men posed in front of family portraits, school-aged children at work at their studies, and pretty young women sporting the latest in fashionable hairstyles. In comparison, Collier's photographs tend to reinforce stereotypes. He often posed old men in such a way as to emphasize their ethnic features.

Roy Stryker, head of Delano and Collier's agency within the FSA, saw his workers as embarking on a documentary exposition and encouraged them to record whatever caught their interest, even if it wasn't directly related to the agency and its agenda. The 1930s saw many manifestations of this documentary spirit as novelists, artists, journalists, social workers, photographers, oral historians, and filmmakers all sought to make visible the trials of ordinary Americans during the Depression. As historian William Stott has argued, the Depression, especially in the years of the Hoover administration, was marked by a profound silence. Americans felt its impact every day, but economic statistics alone could not express the depths of their plight. At the same time, America was awakening to itself as a culture. Conventional definitions of culture drew upon the artistic traditions of Europe and urban society. "Culture" was something for the old world, not the new. But by the 1930s, anthropologists' definition of culture as a society's pattern of organization came into currency. Folk life, rural customs, and ethnic traditions became legitimate subjects of study, and Americans found themselves sitting on a rich cultural legacy just waiting to be discovered.

Delano and Collier's photographs find their inspiration in this "documentary impulse." They represent a discovery of America's diversity and a celebration of its people. Certainly they were motivated by a desire to capture the lives of ordinary Americans and make those people's stories known. But their documentary impulse only went so far: Neither Collier or Delano preserved detailed information about their subjects. The brief captions they supplied consistently provide a location where the photograph was taken. Only rarely do they name a subject, and additional pieces of information are almost non-existent, thus frustrating the efforts of historians such as myself. In captioning these images, I have used the information Collier and Delano supplied, supplementing it with any additional biographical information I could draw from primary source research. If readers of this journal have additional information about the individuals captured in these images, they are invited to contact the Jewish Historical Society of Greater Hartford.

To Learn More

Photographs taken by the FSA/OWI are preserved in the collections of the Library of Congress and can be accessed online at http://memory.loc.gov/ammem/fsahtml/fahome.html.

Oral history interviews with Jack Delano, John Collier, and other members of the FSA/OWI photography division are available in the Smithsonian Archives of American Art and can be accessed online at http://www.aaa.si.edu/collections/oralhistories/transcripts/oralhist.htm.

There are several good histories of the FSA/OWI photography division and of the popularity of documentary work during the 1930s and 1940s. See William Stott, *Documentary Expression and Thirties America*. Chicago: University of Chicago Press, 1973; Carl Fleischhauer and Beverly W. Brannan, eds., *Documenting America, 1935-1943*. Berkley: University of California Press, 1988; Daniel, Pete, Merry Foresta, Maren Stange, and Sally Stein. *Official Images: New Deal Photography*. Washington: Smithsonian Institution Press, 1987.

Not all members of Connecticut's rural Jewish communities were themselves farmers. Leon Broder owned a large feed, lumber and farmers supply store in Colchester. Born in Ekatrinoslav, Russia, Broder became one of the town's leading citizens. In addition to holding several elected positions, Broder founded the Colchester Hebrew School and established the Leon Broder Free Loan Association. During World War I, Broder led Red Cross drives and organized Liberty Loan campaigns.

Jack Delano, photographer
November 1940

Jack Delano took this picture of Mr. and Mrs. Abraham Lapping, Jewish poultry and dairy farmers in Colchester, Connecticut, in 1940. Born in Russia, Mr. Lapping immigrated to the United States in 1913. Making a living as farmers was hard work and the family earned extra money by lodging summer tourists. The Lappings were also longtime members of the Connecticut Milk Producers Association. By posing the couple in front of family portraits and reading a Yiddish newspaper, Delano emphasizes their ethnic identities.

Jack Delano, photographer
November 1940

This photo of Abraham Metzendorf, a poultry farmer in Ledyard, Connecticut, shows the farmer plucking feathers off the bird. Mr. Metzendorf immigrated from Poland and lived in New York City where his children May and Ruth were born before the family moved to the rural life in Connecticut. Delano's caption states that the farmer gets his turkeys ready for Thanksgiving market. Delano took several pictures of Thanksgiving preparations among different ethnic groups, a clear reference to the fact that this immigrant had become part of the American culture.

Jack Delano, photographer
November 1940

John Collier, photographer
August 1942

These two images are typical of the work of John Collier. Neither man is named in the photographer's original record. Instead they are posed in profile to emphasize their ethnic features.

John Collier, photographer
August 1942

Images of village churches were frequent subjects for FSA photographers. This photograph shows a variation on the theme—a small, rural, synagogue in the Huntington District near Newtown, Connecticut. Familiar, yet slightly exotic, images like this suggested the way in which the American experience incorporated immigrant traditions.

Jack Delano, photographer
November 1940

Delano took several pictures of the Hebrew School in Colchester, Connecticut. In this image, two boys study a map of Palestine.

Jack Delano, photographer
November 1940

Taken in 1942, this image of a Jewish rabbi and farmer talking with his neighbor, a boy identified as of German descent, had clear political implications. In contrast to the violence of Hitler's Germany, the two Windsor residents epitomize peace and harmony.

John Collier, photographer
August 1942

Delano took this photograph of Paul Klappersack to illustrate the work of the Farm Security Administration which loaned Klappersack funds to purchase his own poultry truck. By photographing Klappersack from a low angle, Delano fills the frame with the young farmer who appears strong and capable—a sound investment for the government agency.

Jack Delano, photographer
September 1940

FSA photographers often emphasized ethnic features and immigrant traditions, but in this photograph of a Jewish farmer's daughter, Delano shows the process of acculturation. Posed on her knees in the bean patch, the beautiful young woman appears as American as her favorite movie star.

Jack Delano, photographer
September 1940

"It was a completely Orthodox service... And when they built the *shul* [Anshei Israel Synagogue], the men and women still sat in different sections."

Ethel Botnick

FAITH AMIDST THE FIELDS
Connecticut's Country Synagogues

Mary M. Donohue
Photography by Robert Gregson

When we think of the New England countryside, images of stone walls, one-room schoolhouses, and town greens come to mind. But wait—what about that Star of David? In Connecticut, tiny synagogue buildings serve to remind us that a strong Jewish heritage, though nearly hidden from sight, thrived even in the state's remote, rural areas.[1]

Colonial records contain references to Jews in Connecticut as early as the 1660s, but the Jewish population remained small throughout the seventeenth and eighteenth centuries. That was in part because Colonial Connecticut did not allow Jews to worship publicly. As David Dalin and Jonathan Rosenbaum note in *Making a Life, Building a Community: A History of the Jews of Hartford,*[2] Connecticut's Royal Charter of 1662 permitted only the practice of the Christian faith. "Further, it denied Jews the right to assemble as a religious community, build houses of worship, or own property in which to bury their dead," affording these rights only to members of the Congregationalist Church. Even after the American Revolution, Connecticut did not adopt a state constitution and continued to follow the Royal Charter in conducting government affairs. When the new state eventually adopted a constitution in 1818, that document very deliberately did not extend full religious freedom to Jews. Only Christian denominations were allowed full rights to worship.

But the immigration of German Jews to the Hartford and New Haven areas in the 1830s and 1840s soon created a sizable Jewish population in the state. In May 1843, Jewish leaders submitted a petition to the Connecticut General Assembly requesting a change to the state constitution that would allow Jews to worship publicly. Instead of altering the state constitution, the Connecticut General Assembly enacted Chapter XXXIX of the Connecticut General Statutes, which provided "that Jews who may desire to unite and form religious societies, shall have the same rights, powers and privileges which are given to Christians of every denomination…."[3]

Congregations soon formed in Hartford and New Haven, meeting in homes and rented halls. Financially fragile at first, they purchased existing buildings, including former churches, to house their synagogues. Connecticut's first building built specifically as a synagogue, Temple Beth Israel, was erected in 1876 in Hartford on Charter Oak Avenue.

As Yiddish-speaking Jews from Eastern Europe began arriving in the 1880s to Connecticut cities such as Bridgeport, Hartford, and New Haven, they formed their own congregations. *Landsmanshaften*, societies of people who had come from the same European village or area, often were started as a *chevra kadisha*, or burial society. In the first half of the 20th century, Hartford had at least six *landsmanshaften* and 47 Jewish cemeteries.[4] Synagogues became known as the "Hungarian shul" or the "Russian shul." But in the Connecticut countryside, synagogues were built to serve the local farming community and typically served Orthodox congregations. Country synagogues built before 1945 remain today in Columbia, East Haddam, Ellington, Hebron, Lisbon, and Newtown. Two—in Chesterfield and Ellington—are known to have received construction loans from the Baron de Hirsch Fund; undoubtedly others did, too. Some buildings that housed congregations may still be standing but have been converted to other uses. Others are gone, including the earliest known example in Chesterfield, lost to a deliberately set fire in 1975.

Traditional religious practice adheres to many strict requirements and made sustaining a rural congregation challenging. Three of these requirements in particular influenced synagogues. The first of these observances forbids riding on the Sabbath. As country synagogues were often built on farmland donated by one of the congregants, this no-riding rule limited the number of families who could attend services. The second rule requires a *minyan,* a quorum of ten adult male Jews, to establish a congregation and for recitation of certain parts of the service. The third, pursuant to standard Orthodox practice, seats women apart from men. Normally such separate seating is in galleries, but country synagogue buildings were too small to accommodate galleries or balconies, so they created separate sections for women, often using a *mechitzah,* a curtain-like partition to divide the space. Not only were these synagogues too small for galleries, they seldom were large enough to support a full-time rabbi, so services often were led by a member. Sometimes rabbis from New York City would be paid to officiate for High Holiday services in the fall.

Despite these obstacles, farm families living in these rural communities stoutly maintained their Jewish faith. In his essay "Jewish Farmers, Ethnic Identity, and Institutional Americanization in Turn-of-the-Century Connecticut," Richard Moss writes;

> Initially, services were conducted at farmers' houses, such as Beatrice Adams' in Colchester, by a rabbi that had been engaged. By 1918, Colchester's Orthodox synagogue had become a major drawing point for Jewish business owners…Lebanon followed the same pattern: worship was instituted in 1903 and continued for thirty years until the town bequeathed an abandoned schoolhouse for use as a synagogue…These congregations served as the linchpin in the social and religious life of the communities.[5]

Jewish farm families that settled in the Connecticut countryside at the end of the nineteenth century were on their own when it came to constructing a Jewish life for themselves and their families in an overwhelmingly gentile environment. Dr. Kenneth Libo, a historian who was raised on a poultry farm in Lisbon, recalls:

> What kept many Jewish farm families going, even in the toughest of times, was an unshakable determination to survive as a Jewish family—to take the necessary steps and make the necessary sacrifices to stay Jewish. There was never any question. Ours was a thoroughly Jewish household in which the calendar revolved around Rosh Hashanah, Hanukkah, and Pesach (Passover). As a child living close to the soil I felt very close to the Old Testament patriarchs.[6]

In the 1930s and after World War II, Jews again fled Europe, and some found their way to the Connecticut countryside. Their arrival boosted membership in many congregations, prompting a need for larger buildings. In Hebron these historic forces converged and led to construction of a new synagogue, which was built in 1940 for Congregation Agudas Achim.[7]

The photos that follow show country synagogues that are New England on the outside and a little bit of the old country on the inside. Small jewel-like interiors combine American design simplicity with Eastern European ornateness. Plain painted walls do contrast with a gilded ark *(aron kodesh)* surmounted with hand carved Lions of Judah and hung with velvet curtains *(parokhet)*. Once you cross the threshold, the ancient symbols of Judaism-the six-pointed Star of David enclosed in a circle; the Decalogue, the Ten Commandments represented as a double tablet; and the *menorah,* a candelabra reminiscent of those in the Tabernacle and the Temple of Solomon greet you, making it clear that this is no ordinary country chapel.

Connecticut's First Country Synagogue

Hebrew Prayer book from the Chesterfield synagogue printed in Vilna, Russian Empire, 1898-1909.

The prayer book is entirely in Hebrew with only the publisher's name in Russian on the title page.

Luckily for historians, the original congregational ledger book for Connecticut's first country synagogue survives. In the possession of the Savin family, the congregation's ledger, featuring its minutes written in Yiddish, begins in 1892 and covers the next twenty-one years. Recently translated from Yiddish to English, it reveals the importance of the synagogue to its members.[8] Despite financial hardship, members paid dues from the beginning. "Meticulous records were kept of donations received for the privilege of reading the Torah, reserved for members of the congregation…, even though everyone could attend services" relates Chesterfield descendant Nancy Savin. The Society adopted a Constitution in 1894 that clearly outlined required behavior during services and within the synagogue.[9]

A burnt fragment of the *mechitzah*, a cloth curtain that hung from a metal pipe along the sides of the one-room sanctuary, separating the women from the men.

Photos courtesy of Nancy R. Savin

Chesterfield Synagogue, 1892
New England Hebrew Farmers of the Emanuel Society
Routes 85 and 161, Chesterfield, Montville

Financed by the Baron de Hirsch Fund, this building was erected in 1892 by the Ben and Rock Construction Company of New London for $900 and was dedicated on May 8, 1892. The property grew to include a barn for the house of the *shochet* (a ritual slaughterer) and a *mikvah*. It was adjacent to the Creamery Building. Burned by an arsonist in 1975, the site is a designated Connecticut State Archaeological Preserve.[10]

Agudath Achim Synagogue, 1927, enlarged 1951; 503 State Road 87, Chestnut Hill, Columbia

Built by seven or eight of the congregation's farm families, this synagogue is located at a crossroads on land donated by a member. The building, immediately recognizable as a synagogue by the *Magen David* on its façade, was enlarged after World War II to accommodate new immigrants, whose presence doubled the size of the congregation.

Prayer shawls *(tallitot)* featuring the distinctive "thread of blue" stripes are worn by male worshipers during prayer as a commemoration of the kind of wrap worn in biblical times.

Agudath Achim Synagogue, interior. The Torah scrolls were kept behind the *parokhet* (curtain) in the *aron kodesh* (ark). The bimah and the ark are located in an apse at one end of the building. Lions of Judah flank a Decalogue (a double tablet representing the Ten Commandments) over the ark, while the *Ner Tamid*, Eternal Light, is suspended over the *bimah*. These motifs are repeated in the ark curtain. The Decalogue is surmounted by a crown used to represent the crown of the Torah—divine power and inspiration.

Knesseth Israel Synagogue, 1913
Connecticut Jewish Farmers' Association of Ellington
236 Pinney Street, Ellington
Architect: Leon Dobkin

Detail of the *Magen David*, Star of David, over the main entrance.

The well-proportioned thirty-by-forty-foot frame Colonial Revival-style Knesseth Israel Synagogue built in 1913 in Ellington by an Orthodox congregation of farm families features simple white clapboards and a pedimented porch. Local historian Dorothy B. Cohen notes that the plans called for "a…modest wooden structure with two rooms partitioned off by a four foot high wall topped with a row of windows. In keeping with the Orthodox tradition, the room that would contain the ark with the [Torah scrolls] was for the men, and the other was for the women. The women's section would also be used for recreational purposes, and contained a pot-bellied stove."[11] Of the total $1,500 building cost, $100 was donated by Jacob Schiff, a board member of the Jewish Agricultural Society in New York.[12] Knesseth Israel Synagogue was moved from the corner of Abbott and Middle roads to its present location in 1954.

Adath Israel Synagogue photographed in 1940 by Jack Delano for the Farm Security Administration. Note how rural the setting is with rolling fields in the background.

Adath Israel Synagogue, built 1919
111 Huntington Road, Newtown

The only known example of a country synagogue in western Connecticut is Adath Israel Synagogue, built in Newtown in 1919. The land on which it is built was deeded to the congregation in 1914 by Israel Nezvesky for building a synagogue to be called Ohel Moshe. Nezvesky signed the deed in Hebrew; the "signature" of Rose, Israel's wife, appears as an X. The parcel was part of a larger farm owned by Nezvesky. Other Jewish farmers lived along Huntington Road, giving rise to the neighborhood's nickname "Little Palestine." Summer visitors from New York also attended services in this synagogue.[13]

The simple white Colonial Revival-style frame building contained a sanctuary on the upper level and a community room and social hall on the lower. In 2007, the congregation dedicated a new building, leaving this one vacant.

Anshei Israel Synagogue, 1936
142 Newent Road, Lisbon

During the restoration work on the building, bright blue paint remnants were discovered on the window sash. Synagogue historian Erica Myers-Russo states that "this is a color symbolic of God's declaration to Moses that Jews should wear a 'thread of blue' in the fringes of their garments to remind them of the Commandments."

The men sat on long benches and the women sat at a separate table located at the rear of the room. This is out of the ordinary for an Orthodox congregation—usually there would be a wall or divider between men and women.

Anshei Israel Synagogue in Lisbon, located just west of Jewett City, is the archetypal country synagogue. Recently restored by the Lisbon Historical Society, it was built in 1936 by about a dozen families who wished to walk to services in keeping with Orthodox Jewish practice. It is a beautiful, small, frame building that could easily be mistaken for a one-room schoolhouse if not for the *Magen David* affixed to the tower over the entrance. Its simple massing, central pavilion/tower, and eaves are features of the Colonial Revival style. A history of the synagogue written in 2006 richly documents its pivotal place in the lives of three generations of worshippers.[14]

The ark, located at the front of the room, is hung with an ivory-colored velvet curtain embroidered in gold thread and sequins. The top of the ark is surmounted by two gold doves, carved by a member of the congregation, symbolizing peace.

Agudas Achim
Synagogue, 1940
10 Church Street,
Hebron
Architect:
Ira Charles Turshen

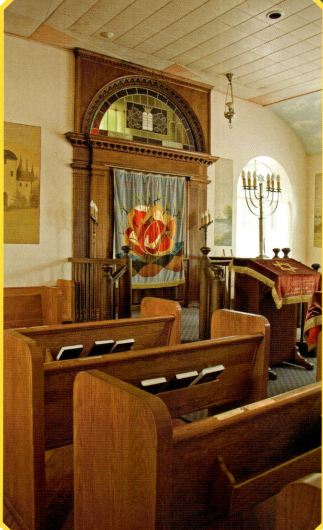

Its late date, use of masonry construction instead of frame, and high style make Agudas Achim in Hebron an unusual country synagogue. Its designer/builder Ira Charles Turshen was a member of the congregation who had a strong interest in design and had briefly attended design school. Turshen was born in a *shtetl*[15] in Minsk, Russia. Coming to America at age five, Turshen's early years were spent in New York City but the family also spent three years on a Colchester farm before returning to New York City. In 1924, he purchased the general store, know as the Amston Grain Mill, in Amston, Connecticut, a few miles from Colchester.

Well-known in the community, Turshen published a local paper entitled *The Amston Poster* and in 1935 the newspaper announced that a committee had formed to erect a new synagogue.[16] Constructed out of bricks from demolished buildings, Turshen's Art Deco building for Agudas Achim is unique among historic Connecticut synagogue buildings. It's stepped massing and planar surfaces make it a good example of Art Deco style. The original murals of landscapes and buildings of the East in the interior are also unusual, perhaps unique.

To Learn More

On Connecticut's country synagogues see:
David F. Ransom, One Hundred Years of Jewish Congregations in Connecticut: An Architectural Survey, 1843-1943, *Connecticut Jewish History,* Volume 2, Number 1, Fall 1991 (West Hartford: Jewish Historical Society of Greater Hartford 1991), 12.

On the Jewish Community of Chesterfield, Connecticut see:
www.newenglandhebrewfarmers.org

Micki Savin, *I Remember Chesterfield, A Memoir.* Bloomington, Indiana: AuthorHouse, 2005.

---

[1] David F. Ransom, One Hundred Years of Jewish Congregations in Connecticut: An Architectural Survey, 1843-1943, *Connecticut Jewish History,* Volume 2, Number 1, Fall 1991 (West Hartford: Jewish Historical Society of Greater Hartford 1991), 12. Connecticut law did not permit synagogues until 1843. In that year, the Connecticut General Assembly enacted Chapter XXXIX of the Connecticut General Statutes, which read, in part, "that Jews who may desire to unite and form religious societies, shall have the same rights, power and privileges which are given to Christians of every denomination." Ransom covers the history and architecture of three types of synagogue buildings built or used between 1843 and 1943: those built as synagogues in urban areas, buildings adaptively reused as synagogues, and small country synagogues. In all, forty-six buildings are inventoried. Since the publication of this CJH journal, several additional synagogue buildings have been surveyed. Historic Resource Inventory forms for the additional buildings are in the collection of the Statewide Historic Resource Inventory of the State Historic Preservation Office of the Connecticut Commission on Culture & Tourism.

[2] David G. Dalin and Jonathan Rosenbaum, *Making a Life, Building a Community: A History of the Jews of Hartford,* (New York: Holmes & Meier, 1997), 6-11.

[3] Dalin and Rosenbaum, 13.

[4] Ibid. 280.

[5] Richard Moss, Jewish Farmers, Ethnic Identity, and Institutional Americanization in Turn-of-the-Century Connecticut, *Connecticut History,* Volume 45, Number 1, Spring 2006 (New Britain: The Association for the Study of Connecticut History, 2006), 47.

[6] Kenneth Libo, correspondence with Mary M. Donohue, 2006.

[7] The primary material for the discussion of the historical and architectural significance of Connecticut's synagogue buildings comes from the survey work by David Ransom commissioned by the State Historic Preservation Office of the Connecticut Historical Commission, now the Connecticut Commission on Culture & Tourism.

[8] The minutes were translated by Mark Nowogrodzki for Nancy R. Savin, Riverdale, New York. The ledger with the minutes is in her possession.

[9] Ibid, unnumbered page.

[10] For more information, see http://newenglandhebrewfarmers.org.

[11] Dorothy B. Cohen, *Ellington: Chronicles of Change* (Town of Ellington: 1987), PP. 142-143; also see http://www.ellingtonshul.org/ourcommunityshistory.htm.

[12] Ransom, p. 64.,

[13] Ibid, 127.

[14] Erica Myers-Russo, *A Refuge in the Country Anshei Israel Synagogue,* (Lisbon: Lisbon Historical Society, 2006).

[15] A *shtetl* is a small town with a large Jewish population in Central and Eastern Europe prior to the Holocaust.

[16] Morris, Lotti Turshen, "Izzy Turshen: Hebron's Renaissance Man", (accessed May 21, 2010) available from www.hebronhistoricalsociety.org.

Left to right: Agudas Achim (interior), Hebron; Agudath Achim, Columbia; Knesseth Israel, Ellington; and Agudas Achim, Hebron

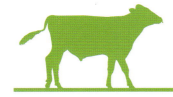

"It seems that in the old days when we moved here there was not enough money coming in from the milk to pay the taxes, to pay the insurance, because these bills come when you own property, so my family started taking boarders in, too. There were two bungalows here that housed about four families...We spent our time as kids sleeping in the barn because our rooms started to be rented."

Marjorie Jaffe Major, Lebanon

THE CATSKILLS OF CONNECTICUT:
Jewish Farming Communities as Summer Retreats

Briann G. Greenfield

For Jewish immigrants who purchased agricultural property in Connecticut's rural towns, the farm was a place of work. Cows had to be milked, land tilled, and crops tended. The day started early with chores and often lasted long into the night. It was a hard life full of physical labor and economic risk. But for another group, urban Jewish workers who escaped the heat of the city to vacation in Connecticut's countryside, the farm became a place of leisure.

The efforts of the Jewish Agricultural Society to resettle Jewish immigrants in rural Connecticut corresponded with a dramatic expansion of vacations as a form of recreation. Before 1910, vacationing was restricted to middle and upper-class whites, those with the resources to escape the work-a-day world, at least for a period of time. For immigrant laborers, the experience of a break from work was more typically associated with seasonal layoffs and unemployment than with rest and recuperation. But by the first decades of the twentieth-century, the labor movement was beginning to secure a living wage for industrial workers, providing them with the means, if not the motivation, to seek time off. At the same time, the twentieth century saw the development of a new culture of leisure, one that cut across lines of class and ethnicity. Short moving picture films screened at Nickelodeon theaters were one of the first forms of commercial entertainment available to workers. Inexpensive and accessible, they nevertheless provided their audiences with a transformative experience, one that lifted the viewers into a new environment. Trips to amusement parks such as those on Coney Island provided a longer respite and also suggested the possibility of travel.

Summer boarders began arriving at Connecticut farms in the nineteen teens, but some of the very first urban summer visitors were in fact workers, organized and transported by the Jewish Agricultural Society to provide much needed labor during the planting and harvesting seasons. The arrangement took advantage of the fact that urban industrial workers often experienced layoffs in the summer, the so-called "slack season." But according to a February 5, 1928 *Hartford Courant* article, the society justified the transportation of workers in more than economic terms, arguing that it would provide "health and vacation" for the travelers.

Whether the workers themselves ever equated the hard, physical labor they experienced on the farm with a vacation is doubtful. Still, two things are clear: Cities were unpleasant places during the summer months. With sweltering temperatures, urban pollution, and frequent outbreaks of disease, urban residents had many reasons to flee the city for rural farms. At the same time, members of the Jewish working class had few options when it came to making their summer plans. Not only were their budgets tight, but anti-Semitism, dietary restrictions, and the desire to maintain their religious practice meant that few vacation destinations were possible. In this regard, the farm filled a significant need. A rented room in the farm house could be had for as little as one dollar a week. Amenities were scarce, often limited to a few hammocks strung out in the orchard to catch cool breezes. But the simplicity of the farm vacation suited a working class clientele that couldn't afford extra expenses. Indeed, many vacationers took advantage of the "*koch-a-lein,*" an arrangement in which guests cooked for themselves, either in the family kitchen or a special summer kitchen built specially to accommodate visitors. The *koch-a-lein* was prized by busy farmers wives as well as budget conscious travelers, eager to economize.

Jewish farmers were not alone in opening up their farms to paying summer guests. Across Connecticut rural farmers made summer boarding an integral part of the diversified farm economy.

Still, the fact that these were Jewish homes connected to larger Jewish communities was important to this particular group of vacationers. Even in the Catskills, the region most famous as a Jewish vacation destination, travelers of the Hebrew faith could face anti-Semitism unless they boarded at a resort catering to a Jewish clientele. Boarding with a Jewish family was, therefore, a way to avoid discrimination. By the nineteen teens, Jewish farm communities were established enough to allow guests to maintain their religious and cultural practices. Vegetables, dairy products and kosher chickens were available on the farms where they boarded. Kosher butchers, country *shuls*, and Yiddish publications could be had in the local community. For Jewish Americans accustomed to an urban environment where their faith and cultural practices flourished, the rural communities of Connecticut's farms provided a practical alternative.

If a farm vacation filled the needs of working class Jewish vacationers, the advantages of taking in summer boarders were even more apparent for the farmers themselves. In a tight economy, farmers looked to make extra money any way they could. Many had been urban residents themselves and therefore had networks of relatives, friends, and old acquaintances who could recommend their boarding services to other city dwellers. Those with prior work experience as peddlers or restaurant staff also had an advantage in that they understood the need to cater to their clients and to provide a level of customer service. For them, the work of running a boarding house was a relatively easy adjustment, even if it meant that their children and other family members might have to vacate their own bedrooms to accommodate additional guests. Still, it is important to recognize that the boarding business was not ideal for the serious farmer. The arrival of summer boarders overlapped with some of the busiest periods in the agricultural cycle. As a *Hartford Courant* article acknowledged, farmers who relied on boarding as part of their income often had to resort to less labor intensive crops

Eventually, many Jewish families gave up the work of farming altogether and turned exclusively to tourism. While individual families still took in boarders, others built dedicated hotels and resorts aimed exclusively at the tourist trade. The process of converting to a tourism economy happened most quickly in the town of Colchester, a community with a long agricultural history but without the rich soils of the Connecticut River Valley towns. By the 1920s, Colchester boasted several full scale hotels and resorts, including the Broadway House, Fairview House, Sultan's Hilltop Lodge, Schwartz's, Kessler's, Horowitz's, and Dember's. According to the 1920 census, Colchester's population numbered 2,050, a number that jumped to over 10,000 during the summer season. Some of the guests came from nearby Hartford and New Haven, but many traveled from New York City. Indeed, up until 1950, a special train ran from Grand Central Station bringing summer tourists to Colchester. The influx of summer residents did much to foster the development of a Jewish business district in the town's center.

Colchester's summer industry was a fleeting one. The Depression of the 1930s forced many resorts to close their doors. When economic prosperity returned, the Jewish resort industry moved to the nearby towns of Moodus, Lebanon, and East Haddam as hotel owners built more modern complexes complete with golf courses, swimming pools, and dance-hall stages where well-known performers such as Eddie Fisher and Milton Berle performed on tour. The postcards that follow document Colchester's resort age.

To Learn More

On the history of summer farm boarders and Connecticut's Jewish resorts see:
Janice P. Cunningham and David F. Ransom, "Back to the Land: Jewish Farms and Resorts in Connecticut, 1890-1945," sponsored by the Connecticut Historical Commission and co-sponsored by the Jewish Historical Society of Greater Hartford, 1998. (Copies may be obtained by contacting the Jewish Historical Society of Greater Hartford)

"Jewish Farmers Prosper in Connecticut," *Hartford Courant*, February 5, 1928.

Additional images of Colchester's resorts can be viewed at the Colchester Historical Society.

On the history of vacations in America see:
Cindy S. Aron, *Working at Play: A History of Vacations in the United States*. Oxford: Oxford University Press, 1999.

Few studies exist on the history of the farm vacation industry. One of the best is Dona Brown's study of the farm vacation industry in Vermont. See Chapter 5 of her book, *Inventing New England: Regional Tourism in the Nineteenth Century*. Washington, DC: Smithsonian Institution Press, 1995.

The following resort postcard images are courtesy of the collections of the Colchester Historical Society.

Colchester's summer visitors arrived via the railroad. By the 1890s, Colchester had its own spur line connecting it to the New York, New Haven and Hartford line, making the rural community easily accessible to New York City Jews.

Elgart's Casino, Colchester, Conn.

Faced with financial problems, the Elgart family briefly operated this summer resort in the late 1910s and early 1920s on land which was previously dedicated to the family's grist mill operation.

JAFFE HOTEL, COLCHESTER, CONN.

This postcard of the Jaffe Hotel shows the kind of rural amenities sought by urban visitors—grassy lawns and ample shade trees.

FAIRVIEW HOTEL, COLCHESTER, CONN.

Located at 119 Broadway, the Fairview Hotel was built in the late nineteenth century as a private home, but by 1867 was being used as a rooming house for mill workers at Colchester's Hayward Rubber Company. In 1919 Harris Cohen transformed the building into a hotel, specializing in the summer trade

Charles Levy, a newcomer to Colchester from New York City, purchased the former Elgart Casino Mill in 1922 and developed it into one of Colchester's largest and most successful resorts. The body of water featured in this postcard once functioned as a mill pond for the Elgart grist mill, but was transformed by Levy into a center for recreation.

Levy's Grand View resort employed the Adirondack style in its fence design and outdoor furniture as a way to connect Colchester to the forested resorts of upstate New York and to suggest a life of rustic leisure.

The annex of Levy's Grand View was decorated with a string of electric lights, a modern spectacle which contrasted sharply with the resort's rural setting.

In an era before air conditioning, wide porches, like those found on Grief's Hotel, provided visitors with a way to catch cool breezes and relax in the shade.

The Pleasant River House shows the simplicity of many early Colchester resorts. The sign in front promises "tourists accommodated" and "meals served."

Max and Mollie Cohen operated the Shady Nook House at 272 Norwich Avenue, near Colchester's center. While many farms took in summer visitors, their success also encouraged the development of in-town boarding houses, including Shady Nook.

Photo Credits

The Jewish Historical Society of Greater Hartford wishes to thank the following individuals and institutions for providing the additional photos used in this volume:

>Colchester Historical Society
>
>Mary M. Donohue
>
>Robert Gregson, Connecticut Commission on Culture & Tourism
>
>New York Public Library
>
>Nancy R. Savin
>
>The exhibit "Geretenish ('Harvest'): Gershon Camassar and the Jewish Agricultural Experience in Eastern Connecticut," presented by the Norwich Cultural Arts and History Project in collaboration with the Norwich Arts Council and the Otis Library, 2009
>
>Lebanon Historical Society

If readers can supply more information on any of the images or subjects in this journal, please contact the Jewish Historical Society of Greater Hartford.

Chart of Interviewees

Interviewees	Dates	Interview Date	Place of Residence
Adler, David	b. 4/18/1915 d. 6/13/2005	2/14/1986	Colchester
Adler, Ruth	b. 4/18/1913 d. 10/9/2004	2/14/1986	Colchester
Agranovitch, Lester	b. 1/29/1915 d. 12/23/2007	Fall 2005	Colchester
Bercowetz, Irving	b. 2/18/1916 d. 2/11/2002	11/10/1980	Bloomfield
Cohen, Rubin	b. 3/25/1911 d. 2/21/1999	2/14/1986	Colchester
Cutler, Belle Bercowetz	b. 6/15/1912 d. 6/20/1999	11/10/1980	Bloomfield
Goldberg, Bernard	b. 6/12/1916 d. 5/16/2008	2/10/1986	Colchester
Himmelstein, Rachel	b. 5/15/1895 d. 3/2/1993	2/14/1986	Colchester
Hopfer, Kurt	b. 5/30/1920	1/1/2006	Norwich
Kiotic, Irving Sol	b. 10/26/1916 d. 1/4/2004	8/28/2003 (1998)	Lebanon
Laufer, Jacob	b. 5/5/1922 d. 10/1986	2/20/1986	Colchester
Laufer, Sarah	b. 12/3/1922 d. 8/20/2004	2/20/1986	Colchester
Liebman, Harold	b. 9/1923	6/20/2005	Lebanon
Major, Marion Jaffe	b. 11/22/1924	5/19/2005	Lebanon
Nassau, Arthur	b. 3/3/1935	Fall 2005	Avon
Polinsky, Harvey	b. 8/20/1938	Fall 2005	Jewett City
Schwartz, Stephen	b. 8/23/1943	2/27/1986	Colchester
Slosberg, Gurdon	b. 1920	6/28/2005	Norwich

The Jewish Historical Society of Greater Hartford, founded in 1971, is a non-profit beneficiary agency of the Jewish Federation of Greater Hartford. The Society's primary function is to collect, organize, preserve, publish and exhibit historical, political, social, and religious documents, photographs, memorabilia and oral histories as they relate to the Jewish community of Greater Hartford. By providing historical information and resources to individuals, educational institutions, and civic and social organizations, the JHSGH hopes to promote historical research and create community awareness and understanding of the growth and development of the Jewish contributions to the Greater Hartford area. The Society's main commitment is to reach the largest audience through exhibitions, publications and educational programming.